AGAINST
THE
FIRES
OF
HELL

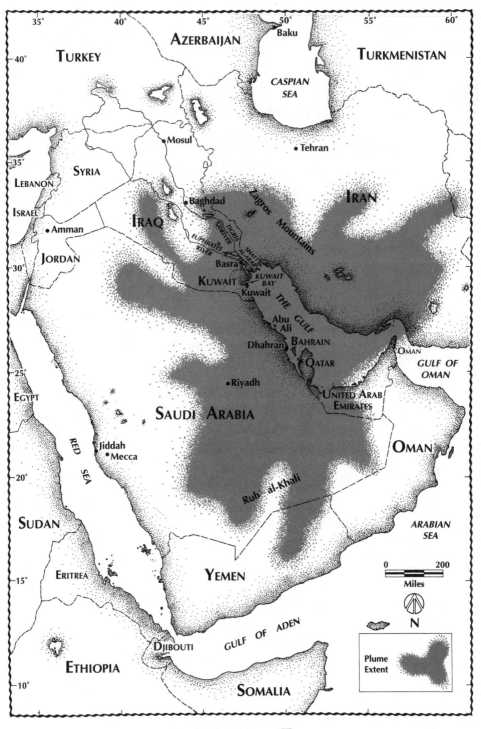

THE MIDDLE EAST

T. M. Hawley

AGAINST
THE
FIRES
OF
HELL

The Environmental Disaster
of the Gulf War

HARCOURT BRACE JOVANOVICH, PUBLISHERS

New York   San Diego   London

Requests for permission to make copies
of any part of the work should be mailed to:
Permissions Department,
Harcourt Brace Jovanovich, Publishers, 8th Floor,
Orlando, Florida 32887.

Library of Congress Cataloging-in-Publication Data
Hawley, T. M.
Against the fires of hell: the environmental disaster of the Gulf
War/by T.M. Hawley.—1st ed.
p.   cm.
Includes bibliographical references and index.
ISBN 0-15-103969-0
1. Persian Gulf War, 1991—Destruction and pillage—Kuwait.
2. Persian Gulf War, 1991—Destruction and pillage—Persian Gulf
region.   3. War—Environmental aspects—Kuwait.   4. War—
Environmental aspects—Persian Gulf Region.   5. Oil wells—Fires
and fire prevention—Environmental aspects—Kuwait.   6. Oil wells—
Fires and fire prevention—Environmental aspects—Persian Gulf
Region.   I. Title.
DS79.744.E58H38   1992
956.704'3—dc20   92-21011

Designed by Trina Stahl

Printed in the United States of America

Maps drafted by E. Paul Oberlander

First edition

A B C D E

Acknowledgments

$\text{\large M}$ORE THAN ANY other person, my wife Liz inspired and supported my work on this book. Her enthusiasm helped to sharpen my focus and allay irritations along the way. My friend Kevin Galie offered assistance of many kinds. Two friends from college days, Ross Schennum and George Stubbs, went over the chapters as I wrote them.

William H. MacLeish and Paul R. Ryan—two former editors of *Oceanus* magazine at Woods Hole Oceanographic Institution—and H. G. Bissinger gave me much-needed advice. My agent, Laura J. Blake, and my editor, Anne Freedgood, believed in the project from the beginning and supported me throughout the writing. Roslyn Schloss gave the manuscript a final, professional going-over. Vaughn Andrews, Josh Getzler, Marianna Lee, Trina Stahl, and others whom I have not even met all worked hard on the book.

Thanks to the openness the fire fighters in Kuwait and the cleanup workers in Saudi Arabia displayed toward reporters, the world has a better view of how our civilization deals with the grossest side effects of its petroleum dependency. T. B. O'Brien of O'Brien Goins Simpson consistently kept me up to date on the fire fighters' progress as 1991 burned its way into history. Joe Bowden of Wild Well Control, Raymond Henry of Red Adair Co., Mike Miller of Safety Boss, and Coots Matthews of Boots & Coots were all patient with me as I tried to understand how they managed to be so successful under such incredible circumstances. Stacy Miller of Safety Boss and Joetta Janczak of Red Adair Co. provided essential help in arranging interviews with their companies' fire fighters and in supplying various bits of information. The Kuwait Oil Company, the corporate entity with the most to lose from the oil fires, took pains to ensure that writers like myself got the most thorough possible view of the situation. My especial thanks to Sa'ad al-Mousa in the Engineering Group and Ali Murad in the Public Information Office.

Khaled al-Razni of the Kuwait Ministry of Information helped me orient myself and set up interviews. Mamoud Abdulrahim, Ibrahim Hadi, and Moustafa Dessouki of the Ministry of Public Heath were generous with their time. Special thanks to Sami al-Yakoob of the Kuwait Institute of Scientific Research who invited me to his home and diligently arranged meetings with his fellow scientists, who were among the first to supply the outside world with information on the devastating pall of smoke and petroleum fog that obscured Kuwait for most of 1991. Those scientists—Fatima Abdali, Jassem al-Hassan, and Ali Khuraibet—were refreshingly open with their findings and opinions on the crisis.

In Saudi Arabia, the personnel of the Meteorology and Environmental Protection Administration, by seeking out the world's best expertise in oil-pollution cleanup, hastened the process of admitting the foreign experts into their relatively closed country. MEPA's vice president, Nizar Tawfiq, graciously arranged a week of interviews and site visits for me, with Abdul-Jaleel al-Ashi, an MEPA official, as my guide. I am grateful to Abdul-Jaleel for not only his generous hospitality but also his thoughtful remarks on the spill, the religion of Islam, and Saudi society. David Olsen of MEPA was among my very first contacts in the Middle East. Without his help, my task would have been far more laborious; in addition, he reviewed the manuscript for accuracy and completeness.

Several journalists who covered the crisis shared with me their sources and crucial bits of information they uncovered. My especial thanks to John Horgan of *Scientific American*, Tony Horwitz of the *Wall Street Journal*, and Matthew Wald of *The New York Times*. Both Tony and Matthew also provided me with trenchant anecdotes from Kuwait in the days just after liberation from the Iraqis.

Farouk El-Baz of Boston University's Center for Remote Sensing, John Evans of the Harvard School of Public Health, and Robert M. Russell of the Tufts University Human Nutrition Research Center each provided a wealth of information and heartfelt encouragement throughout the creation of this book. Their concern for a reliable account of the war's ecological aftermath extended to reviewing manuscript chapters for accuracy and completeness. Other scientists who contributed information in the form of interviews, research articles, and/or critiques of the manuscript include Sylvia Earle of the U.S. National Oceanic and Atmospheric Administration; Peter V. Hobbs of the University of Washington; Paul Horsman of Greenpeace U.K.; Sanjay S. Limaye of the University of Wisconsin-Madison; Douglas H. Lowenthal of the Desert Research Institute; John Robinson of the U.S. National Oceanic and Atmospheric

Administration; Richard D. Small of Pacific-Sierra Research Corporation; Abdullah Toukan, science adviser to King Hussein of Jordan; and Richard P. Turco of the University of California, Los Angeles. Although all of them patiently answered my questions and reviewed my interpretation of their answers, any mistakes remaining in the book are mine alone.

Finally, a host of friends encouraged my work and stimulated my thinking. They include my parents, Tom and Bernice Hawley; Monroe and Nora Johnson; Dan and Kitty Beller-McKenna; M. Cathleen Krebs, and the rest of the St. Paul Adult Choir in Cambridge, Massachusetts; Bernard and Mary Edstrom; and members of the G. I. Nutrition Laboratory at Tufts University Human Nutrition Research Center.

To my wife, Liz,
and my sons,
Pat, Kevin, and Ben

Contents

Prologue:
Against the Fires of Hell

❮ ❮ ❮ ❮ ❮ ❮ ❮ M ARCH 3, 1991, AP-
proaching Kuwait International Airport from the south,
blazing sun above, a swirling ocean of black fog below, the
plane began what each of its eight passengers later described
as a descent into hell. From the al-Maqwa oil field just south
of the airport, and on to the south for as far as the eye could
see, the sands of ash-Shaq'q desert were obscured by the
slowly undulating blanket of dark smoke. Held aloft by currents of air, it was a blanket that a jumbo jet could fly
through, but it was capable of coating trees with a layer of
tar and of smothering the lungs of birds, camels, and sheep,
and it would eventually reveal stark political divisions within
a society that previously seemed united by its inheritance of
the source of that choking mantle—crude oil. Off to the
north, on the other side of Kuwait Bay, another gigantic

plume of smoke and petroleum droplets billowed upward and spread downwind from the ruins of two other oil fields.

The only sign of land beneath the smoke was hundreds of fiery furnaces ablaze in every configuration imaginable. Some were single jets of pitch and flame erupting straight up from the desert floor; others, enormous howling blow-torches with flames shooting off in several directions at once. There were also rivers of flame that flowed into ever-growing pools and lakes of crude oil. Burning lagoons spilled over the roads in the Greater Burgan oil field of southern Kuwait, igniting both storage tanks and the pipelines that linked the oil field to the Kuwait Oil Company's refining and export facilities at Mina Ahmadi.

From their seats in the aircraft as it began its final approach to the airport, the passengers got their first whiff of the fumes: a stinging, acrid smell that not only irritated their nasal membranes but stuck to their skin, hair, and clothing. Like the million or so inhabitants of Kuwait during 1991, they will always remember that odor and taste, because for the next eight months they would struggle to put out those fires. It was a vision of hell, indeed, that they were peering down on. But they were studying it as surgeons might study the victim of multiple wounds.

The passengers included Raymond Henry, an associate of the legendary Red Adair. Accompanying Henry were Joe Bowden and Boots Hansen, both of whom got their starts fighting oil-well fires in Adair's shadow. Now, however, they were all in business for themselves, with Boots now a partner with Coots Matthews and Bowden as president of his own company. Also on board was Paul King, an Oklahoman who had worked in Kuwait for the Kuwait Oil Company for about fifteen years. This was his first return since he had fortuitously left on vacation the previous July, about a week before the Iraqi troops arrived. His job was to acquire, and arrange for the delivery of, every piece of equipment that the fire fighters would need in their effort. He figured that it would take at least two years to put the fires out. Testifying before a U.S. Senate Committee three months after the first fire fighters arrived in Kuwait, Adair cited delays in the delivery of heavy equipment and said that if the delays were not eliminated quickly, five years might elapse before the last fire would be snuffed out. He called an official Kuwaiti timetable that allotted nine to twelve months for completion of the task "a bunch of malarkey."

As it turned out, nine months was more time than the fire fighters needed. King didn't realize how large his logistics operation would grow, and Adair hadn't counted on the dozens of crews from companies all over the world who would join the effort, or foreseen the efficiency that would come from killing so many wells at once.

The men on the flight, the first into Kuwait after the ceasefire ending the Gulf War, had all seen the televised version of the war, the version in which laser-guided "smart" bombs and other sophisticated weaponry destroyed Iraq's command-and-control facilities with antiseptic precision. Now they were seeing, inhaling, and tasting the other side of the war's destruction, environmental carnage that at once evoked Julius Caesar's poisoning the soil of Carthage with salt and spotlighted our industrial civilization's dependence on energy from petroleum. An old saying has it that wars determine not who is right, only who is left, but the victors who were left in Kuwait at the end of the Gulf War didn't have much of a country now. For this conflict, which was much heralded as the beginning of a new era in high-tech war engines, will probably be remembered longer for the unimaginable degree to which one of the belligerents employed the most ancient of war strategies—scorched earth. From Caesar's campaigns in Africa, to the burning of Moscow in September 1812, and to the use of Agent Orange in Vietnam, armies in retreat and armies on the march have always laid waste to the environment in order to deny their adversaries the bounty of the land being fought over.

Henry and the others were among the first eyewitnesses of the new standard in scorched-earth policy set by the Gulf War. But the fires in Kuwait were merely the most obvious example of the war's environmental horror: they only consumed about a third of what came roaring out of the ground. The smoke from the largest conflagration in history finally cleared in November 1991 to reveal a landscape in ruins. Weathered crude oil equivalent to more than a month's supply of refined petroleum products in the United States had accumulated in miasmic swamps and lakes covering hundreds of square miles. The oil constituted about half the contents of those lakes; the other half was briny water that rose with it from the depths. With each passing day, the oil became thicker and finally congealed as its lighter components evaporated. And it was a thick goo to begin with, since many of its flammable components had already gone up in smoke. The process is still continuing in 1992, and the thicker the oil gets, the tougher it will be to pump it out of the lakes. The lakes will almost certainly never be completely drained, and cleaning the soil contaminated with oil and smoke is practically out of the question.

An even deadlier problem is soil contamination from land mines and unexploded ordnance. No effort to rid the land of the war's legacy of Iraqi mines and of "rock eyes"—the still-lethal remnants of Allied cluster bombs—will ever be complete. Fearing death or injury, the Kuwaitis have abandoned their tradition of springtime encampments in the desert.

Beyond the vast areas covered by oil lakes and war matériel lie the expanses of desert savaged by months of fallout from the fires. The flat landscape of Kuwait and northern Saudi Arabia, once the light tan color of a lion, is now stained black under millions of tons of oily soot that contains the myriad chemicals present in crude oil and its combustion products. The sweeping wilderness covered by the fallout is far too vast ever to be cleaned by man, but in time the wind-blown sands of the desert will slowly overspread the enormous stain of war, burying the evidence but not the memory of the most terrifying air-pollution catastrophe in history.

The survivors of the pollution will have long been gone. After enduring occupation by hostile forces while the hereditary rulers of their land fled and lived abroad in luxury, after having had their property confiscated, their loved ones abducted, sometimes even tortured or killed, those survivors had looked forward eagerly to the rebuilding of their country and its society, even if the work needed to be carried out beneath black skies and amid noxious air. Among the survivors were scientists who, during the occupation, had clandestinely educated their compatriots in methods of avoiding the agents of chemical and biological warfare; why would they not be able to devise precautions against whatever dangers were present in the smoke from the oil fields? How could smoke from crude-oil fires, even in unprecedented quantity, be a more serious threat than what they had suffered during the occupation? All that was necessary, they believed, was an aggressive public-education campaign, giving the population easy-to-follow steps that would minimize their exposure to the riskiest chemicals in the complex cloud that blackened the noonday skies. Instead, once the American-led forces made Kuwait safe for the return of the government, its officials loudly proclaimed that they had found remarkably low levels of the few pollutants they chose to study and ignored the arguments of those who called for precautionary measures to shield the population from dangerous pollutants that were probably in the cloud, but for which the cloud was not being analyzed.

Weeks before Iraq's soldiers detonated their first oil well in Kuwait, they had sent a deluge of crude oil into the Gulf.

From January to June 1991, the largest oil spill ever—aside from the accumulated releases on the land in Kuwait—the equivalent of between twenty-five and forty *Exxon Valdez* spills, bore down on the Gulf coast of northern Saudi Arabia. The same stretches of coastline had only recently begun to recover from the approximately two million barrels of crude oil that spewed from Iranian offshore platforms after Iraqi missile attacks in 1983. Although that spill was, until Operation Desert Storm, the second largest in history, it was only about a fourth of the size of the behemoth unleashed during the Gulf War, and ten months passed as it flowed into the Gulf at an average rate of 7,000 barrels a day. In grim contrast, the bulk of the Gulf War spill gushed into the shallow waters of the northern and eastern Gulf over the course of just a few weeks, at an average rate of at least 150,000 barrels a day.

Luckily for the creatures living in the Gulf's deeper waters, crude oil is not highly toxic, as pollutants go. Nevertheless, it contains a host of chemicals that cause changes in the immune systems of fish and that can lead to cancer. The devastating power of crude oil spilled into coastal waters comes principally from its sheer gooeyness. In sufficient quantity, it smothers every living thing caught in its sticky grasp. After it sinks into coastal sediments, binding innumerable grains of sand and tiny creatures into a gritty black mess in the process, it is notoriously difficult to remove.

Although oil continued to leach out of contaminated coastal sediments and into the Gulf for more than a year after the war, essentially all of the floating oil had been removed by the end of July 1991. Using every mechanical method known in oil-spill cleanups and almost religiously avoiding chemical dispersants or biological agents, a handful of cleanup companies managed to recover more than 20 percent of the oil, an unprecedented proportion for such an enormous spill and in absolute terms almost six times the total spillage from the *Exxon Valdez*. But despite the success of the cleanup crews, millions of barrels of oil remain trapped along the Gulf's extremely wide intertidal zone and shallow near-shore waters. Because pristine beaches and salt marshes are of minor economic importance in this area, the prospects are slim for an aggressive shoreline cleanup like the one carried out in Alaska in the years since the *Exxon Valdez* disaster.

The most desperate suffering in the wake of the Gulf War is, however, in Iraq. Especially among the Shiite population of the south, preventable infectious diseases have become endemic, and starvation haunts a land whose children are already weakened and stunted by food shortages brought on by nearly ten years of war with Iran. They saw their fathers and older brothers rise up against the tyranny of Saddam Hussein in the days following the ceasefire, only to be put down with crushing swiftness. Not only were homes destroyed and families slaughtered, but the electrical power plants were in shambles as a result of air attacks during the Gulf War. Without electricity, waste-water pumps ceased to function, and the streets ran with raw sewage as summer arrived with its sweltering temperatures.

Eager to assert his leadership among his own tribesmen in the areas around Baghdad and to punish those in Iraq who dared to defy him, Saddam Hussein managed to repair much of the war's damage to Baghdad's roads and electricity grid. The export of oil, transported in tank trucks along the highway between Baghdad and Amman, Jordan, provided the means to repair the facilities critical to those who sided with him during the brief rebellions after the war. By May, food was plentiful, if expensive, and power was on in Baghdad. But little if any of the food found its way into the Shiite region of the south, perhaps, it has been suggested, because Hussein hoped the suffering of Iraqi civilians would force the United Nations to lift the sanctions against his country.

The indisputable fact is that emaciated infants and children have been dying every day for lack not only of food but also of simple medical supplies that could have kept them alive.

❦ ❦ ❦

THE OIL SPILLS into the Gulf would have stood as the worst incident of oil pollution ever, had the oil fires in Kuwait not overshadowed them. Taken together, the two represent an environmental catastrophe on the order of the explosion at the Chernobyl nuclear-power station or the lethal release of methyl isocyanate at Union Carbide's chemical plant in Bhopal, India. Although the oil that spilled and burned in Kuwait did not suddenly and horribly take human lives in the way that the Chernobyl and Bhopal catastrophes did, the Gulf War's legacy of pollution has been felt the world over and has left millions in the Middle East wondering whether their air is fit to breathe, their water fit to drink, their food fit to eat. The war has transformed these people into involuntary guinea pigs, part of a vast experiment in environmental degradation.

In the industrialized world, news media brought to millions of people various aspects of the fantastic effort to extinguish the oil-well fires. But the fire fighters, despite their success, could not limit the pollution's severity in the Gulf region or halt its dispersion to distant continents. They could bring an end only to what one might call the acute phase of the war's environmental aftermath. The environmental devastation in the Gulf region will continue for decades, and I have devoted most of this book to the challenge to assess and overcome it. The heroic story of the fire fighters merely sets the stage for the larger tale.

Many of the people that I met in the course of researching this book told me that the most important thing for them, as they dealt with the unprecedented catastrophe, was the hope that through their work what had appeared to be an unmitigated disaster could be partially redeemed by the lesson that it taught. I initially met most of them by telephone as I covered the war's environmental angle, beginning with the first predictions of what such a war could mean in an ecological context. After Saddam Hussein opened the floodgates at the Sea Island Terminal and ignited Kuwait's oil wells, I was able to further my acquaintances by attending conferences and seminars on the postwar environment. Finally, I traveled to the Middle East just as the fire fighters were putting out the last oil-well fire, and I met in person many of those with whom I had been conversing over the phone for months. It was a good time for a reporter to be in the region, as people there were beginning to discern patterns that had eluded them while the area was in the grip of the worst part of the crisis. Some said that they would recount the events of 1991 to their children and grandchildren as an example of where violent resolution of conflict leads. Others spoke of an optimism that bound the cleanup workers together: "We had to show Saddam that his hate could not defeat us." Still others told me that, had it not been for this disaster, they would never have believed so many people would come together and work so hard to protect the planet Earth. I have tried to write this book in that spirit.

CHAPTER ONE

The Fog of Peace

≋ ≋ ≋ ≋ ≋ ≋ ≋ D URING THE first few weeks after the fire fighters' vanguard arrived in Kuwait, they walked about in disbelief of the catastrophe it was their job to halt. No amount of video footage or other reports from the apocalyptic wasteland of postwar Kuwait could prepare them for the all-pervading blackness and the sight of hundreds of blown-up oil wells. Smoke and fire stretched to the horizon in every direction. The roar and smell of burning crude oil were their last sensations before falling asleep and their first sensations on waking. It was a different world.

With no tools other than standard oil-field wrenches, they ventured out into the oil fields, tightening bolts on the few wells that were leaking only small amounts of oil and not on fire. Paul King, in charge of supplying the fire fighters, had nothing else to supply them with. Iraqi troops had made off

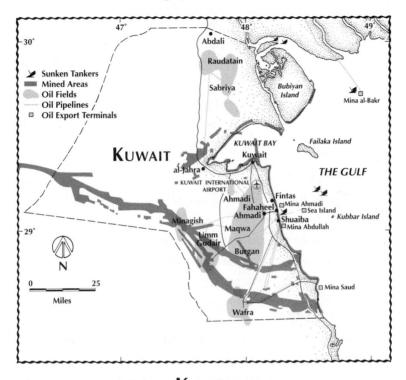

KUWAIT

with or destroyed practically every piece of oil-field equipment in Kuwait. The town of Ahmadi, where the Kuwait Oil Company had its headquarters, was a shambles. King didn't even have water to cook or wash with, much less to satisfy the needs of an army of fire fighters and maintenance workers. Sporadic battles with recalcitrant Iraqi troops still broke out. Kuwait International Airport was closed more ten than not, as mine-clearing crews were just starting their sweeps of the runways, and billowing clouds of acrid smoke obscured vision.

While King exhausted himself trying to rustle up the rudiments of what would become the largest fire-fighting operation in history, Raymond Henry and the other most experienced fire fighters began studying the situation, trying to come up with some ideas of how to make the initial, probing attacks against the pitch, flame, and smoke swirling across Kuwait's oil fields. From one hundred feet up in a helicopter, Henry could hardly believe the views: jets of flame shooting twice as high as he was, belching thick clouds of black smoke and a mist of unburned oil, or billows of white smoke—or was it steam, or a cloud of salt carried up from the depths with the oil? For now, it was just something else to avoid. Henry and the other passengers choppered their way through the al-Maqwa region of the Greater Burgan oil field, carefully staying upwind of the choking smoke and fumes. Suddenly, a great fireball, shining like a miniature sun, floated majestically up through the white plume that rose from the flaming gusher off to the right of the aircraft.

The straight shooters, Henry said to his companions, wouldn't be hard to kill. But getting to them was another matter altogether. It would require building roads across the oil lakes, as well as dikes to keep the lakes from overflowing into the men's operations. Good wind forecasts would be needed too, to help the fire fighters keep out of the smoke as much as possible.

Unburned oil from the hundreds of exploded wellheads had pooled and then flowed along the barely perceptible inclines of the Greater Burgan field, which covers much of the desert of southern Kuwait. The rivulets of crude oil combined and grew into lakes a mile or more long and several feet deep. Especially in the Burgan, and the Sabriya field to the north of Kuwait Bay, the lakes ran together and coalesced into marshes. As they grew and swamped the fiery geysers, flames, greasy with oil fumes, rose into the air above. A lake might writhe in flame for a few hours before it exhausted its oxygen supply, dying from the inside out, but with the wind constantly blowing across the oil, pulled by the gusts of fire surging from the wellheads, a mile-long lake could reignite without warning, suddenly erupting into gigantic flames.

When the wind changed, the oil swamps streamed off in new directions, spreading over service roads, covering mine fields and the ubiquitous shards of deadly matériel strewn across the desert. Driven from an area by the wind, the lakes left behind shining blankets of semiliquid asphalt and spilled into the dug-out fortifications from which Iraqi soldiers had emerged and surrendered just a few weeks earlier. Bunkers still plentifully stocked with artillery shells and other high explosives were lost in a tarry soup of crude oil. There was no way of killing oil wells until some means could be found to cross safely the ever-changing morass of stinking black goo.

Until the fire fighters got close to the wells, right under the screaming 2,000-degree flames, they had only a vague idea of what it was going to take to snuff them out. Water, of course, an ocean of it, and bulldozers, cranes, water cannons, all girded with corrugated tin against the heat, and plastic explosive, which, fortunately, was plentiful in the aftermath of the war. But what if they couldn't put out many of the fires without first drilling relief wells? Some of the wellheads were spraying oil and flames in three or four directions at once, some were obviously oozing oil up to the surface from ruptures sustained who knew how far down. Hoping that the Iraqi explosives had not wrecked the well casings at great depth, the fire fighters would first try to dig the monsters out; if they found undamaged casing before a hole got too deep to work in, they wouldn't have to bother with a relief well. But in the beginning, no one knew how deep the damage went.

For the first two months, no one could even count the number of wells gushing oil in Kuwait, so dense was the filth they were spewing. On February 22, military sources said that "about 140" were burning out of control. Three days later, the estimate had risen to 517, and smoke was already obscuring Iran's Gulf coast. By the end of March, the best estimate was about 600. It wasn't until May, after a hundred wells had already been capped, that the Kuwait Oil Company announced the definitive figure of 732 blown wells. The company figured at that time that the wells were burning as much as 6 million barrels a day, 10 percent of the entire world's daily production. At $19 a barrel, that was nearly $1,500 a second going up in smoke. Even so, Saddam Hussein's act of "environmental terrorism," as journalists had dubbed it, fell short of the rate at which America's motor vehicles were consuming gasoline—7,047,000 barrels a day.

Lying at the center of our planet's longest expanse of desert, which extends from Morocco to central China, Kuwait is a country almost devoid of topography. Not even a well-defined watershed exists there. Aside from Kuwait City, the town of Ahmadi about twenty miles to the south, and the oil refining and exporting towns along the coast to the east of Ahmadi, the only remarkable features visible from the air before the Iraqi invasion were the Wadi al-Batin, which defines Kuwait's western border with Iraq, and pipelines linking the oil fields to the refineries. Within that monotonous expanse, frequent sandstorms, which can last for several days at a time, are responsible for Kuwait's distinction as the country with the highest average amount of inhalable particulate matter in its air.

Despite the seemingly blank slate of the Kuwaiti landscape, the oil industry had relatively little effect on its appearance. The popular image of scores of drilling platforms, or derricks, was nowhere in sight once the wells were completed. A typical oil well in Kuwait is little more obtrusive than a traffic light surrounded by a chain-link fence. The top of a well, known as the "Christmas tree" because of its bristling array of flanges and valves, has all the features necessary to control the flow from the well. Within the well itself, a variable number of concentric tubes reach down to different depths, with the outermost tube, or casing, protecting the subterranean environment and the oil well from each other, and the innermost tube serving as the conduit for oil rising to the surface. Most of Kuwait's wells produce oil by means of underground pressure, which eliminates the need for pumps on the surface but underscores the importance of a device known as a blow-out preventer. Petroleum engineers designed the devices to slam the well shut in the event of a dangerous rise in subterranean pressure. When Iraq's soldiers detonated the explosives around most of Kuwait's oil wells, they destroyed the Christmas trees and blow-out preventers, and did varying amounts of damage to casings and the other tubing near the top of the wells.

Little more than a month after Iraq invaded Kuwait, Paul King and other representatives of the Kuwait Oil Company began meeting with the fire-fighting companies in Tulsa and Houston to plan how to supply the postwar reconstruction effort. He envisioned a hundred fires, a figure that reflected a belief commonly held among the exiled Kuwait Oil Company community that Saddam Hussein's threat to annihilate Kuwait's oil industry was largely bluster. There was also the sheer impracticality of stockpiling equipment for an emergency of unprecedented but indeterminate scope. If King had at that time had the authority to plan for an operation of the size that it eventually grew to, he would have been ordering the special construction of all kinds of heavy equipment. As it was, he had his hands full arranging for the delivery to the Middle East of some of the equipment that would surely help the fire fighters in their task. By the beginning of March he had amassed capping assemblies—which are a kind of Christmas tree—blowout preventers, and earth-moving equipment in Saudi Arabia and Bahrain. But he thought it would be irresponsible to purchase all the bulldozers in the world when their ultimate use was so uncertain. Until he arrived back in Kuwait at the beginning of March and saw what he was up against, his strategy for supplying the fire fighters remained largely an academic exercise.

On returning to Kuwait, King moved into the Kuwait Plaza hotel, which had escaped the looting and arson that Iraqi troops had indulged in at more prestigious establishments, like the Hilton Intercontinental and Meridien. He operated out of the Plaza for a week, ferrying everything that arrived for him at the Kuwait airport to one of the guest houses of the Kuwait Oil Company in Ahmadi. As soon as a runway had been cleared of wreckage and hidden explosives, the planes began bringing in the goods whenever the smoke wasn't too thick overhead. King started from scratch; the Iraqis had looted the company compound of everything from beds to computers. A small desalination plant aboard a Russian ship moored in waters still prickling with mines supplied his drinking water.

Until the British army's Royal Ordnance division established its mine-clearing routine, King, like virtually everyone else in Kuwait, was on EOD, or explosive ordnance disposal, duty. In the first weeks after liberation, Kuwait's hospitals received a stream of civilians maimed by the war's leftover explosives. Victims of postwar score settling and random gunfire added to the toll—King himself was fired at on one occasion. Doctors also saw many people with respiratory and skin complaints.

All though March and into April, the fire fighters struggled to get going. The three Houston-based companies—Red Adair, Boots & Coots, and Wild Well Control—had two crews apiece in Kuwait by March 20. By April, working mostly with equipment scavenged from war wreckage, the five- or six-man crews had brought under control twenty wells, most of which, having been only slightly damaged, required merely the closing of a valve. There were few wells, however, with such minor damage; the Iraqi troops had packed between 300 and 400 pounds of plastic explosive around most of the wells and piled sandbags on top in an attempt to direct the explosions downward and inward, where they would annihilate control valves.

Until a regular flow of heavy equipment started moving into Kuwait, the fire fighters could attack only the "squirters," which yielded to the relatively gentle force of standard wrenches. Normally, when a crew goes to kill a blowout, it uses the tools already on site. But the Iraqis had looted Kuwait's oil fields of every scrap. And there was no water. The residents of Kuwait City were getting by on water carried in tank trucks that contained more than just traces of their previous cargo, diesel fuel. With the Gulf as the nearest source of surface water, and with no way, in the first several weeks, to deliver that water to the oil fields, the fire fighters also had to rely on tank trucks. On several occasions, these supplies gave out just when the crew was gaining the upper hand in an hours-long battle to approach a flaming well.

By the end of March, King had managed to clear out five guest houses in Ahmadi, room for about two hundred men, and had two cargo planes a day carrying food and equipment to the slowly growing camp of fire fighters, bulldozer drivers, welders, and office and domestic assistants. Some of the heavy equipment from Saudi Arabia and Bahrain came in on the flights, but for the effort really to make progress he needed enough equipment to supply over a dozen crews— more of the gargantuan backhoes that could excavate around the worst of the damaged wells and more road-building equipment. The D-8 Caterpillar tractor was his personal favorite among the latter, but there were only twenty-five available in the world. Eight months later, when the last of the fires was finally out, the Kuwait Oil Company owned eight hundred of them.

In Ahmadi, welding crews, composed mostly of Filipinos, Pakistanis, Sri Lankans, and other south Asians who

had survived the Iraqi occupation, worked night and day—
not that they could tell the difference by looking at the sky—
attaching sheets of corrugated tin to the bulldozers and
backhoes that had either been scavenged off the battlefield
or were trickling in on the daily cargo flights. Every piece
of heavy equipment used near a blazing well was armored
with tin to prevent the heat from igniting the fuel in its fuel
tank, a danger further lessened by constant sprays from water
cannons themselves encased in huts of tin. By the time a
long-reach backhoe was ready to go into action against the
blowtorches, sheets of corrugated tin had transformed it into
a kind of mechanical brontosaurus.

Without the airlifted cranes and backhoes, the fire fighters would have been at a standstill, but they still needed
shiploads of material, and the military didn't declare the waters
near Kuwait safe from mines until the latter part of April.
Overland transport was out of the question until roads were
rebuilt—a task itself requiring the very equipment that was
in such short supply. The opportunities for profit were not
lost on customs officers assigned to clear the equipment
through. At every border, delays and costs increased, and to
top it off, Kuwaiti officials effectively grounded the fire
fighters for a few days by demanding an import duty on fuel
brought in from Saudi Arabia—this while more than a billion dollars a day of crude oil was burning and spilling over
the land.

For a while, relations between the fire-fighting crews and
the Kuwait Oil Company were touchy. The whole world
was watching, and worrying about, the fires' effects on
monsoons in Asia and on the climates of distant regions, and
progress against the flames was agonizingly slow. Fire fighters claimed they encountered a stubbornness on the part of
company officials to authorize purchases of necessary equipment. Kuwait's minister of oil at the time, Rashid al-Amiri,
countered by pointing out that the fire fighters' contracts

called for payment on a per-diem rather than a per-well basis, and there were no bonuses for fast work. "Was it not in the fire fighters' interest," al-Amiri reasoned, "for them to work slower rather than faster?"

Red Adair's company and the other two companies from Houston were the only ones in the world specializing exclusively in killing wild oil wells when Saddam Hussein invaded Kuwait. But Adair is quick to point out that it was two other Houston men, H. L. Patton and Adair's mentor, Myron Kinley, who in the 1920s and 1930s pioneered the method of killing wild wells by dousing them with jets of water and suffocating them with blasts of high explosive.

Adair has more than once taken an unplanned ride on gushing oil to the upper reaches of a drilling platform, but he doesn't take risks that he deems unnecessary. He refused the job of capping the blown-out Iranian wells that had been the target of Iraqi missile attacks in 1983, because he could not obtain guarantees of safety for his crews from the belligerents. Almost a decade later, he wasn't about to risk the lives of his men in the mine fields of Kuwait. But the mine-clearing crews could not come close to the burning wells without the kind of protection from the heat that the fire fighters had developed. The parties agreed that Royal Ordnance would clear areas up to about a third of a mile from burning wells. The fire fighters would be on their own from there on in.

Throughout much of March, while Coalition troops drove the Iraqi army back to Baghdad, and the gruesome scene on the Kuwait to Basra "highway of death" played across television screens and magazine pages, the skies grew steadily darker over Kuwait, and sooty air could be seen and smelled a thousand miles downwind. Saudi Arabia's environmental agency had to cancel flights intended to reconnoiter the position of millions of barrels of oil bearing down on the country's coast. The governor of Adana, in southwestern Turkey, forbade citizens to use rainwater blackened by soot for either themselves or their animals. Horses, cattle, camels, sheep, and goats in Kuwait succumbed to poisoned air and oiled fodder, their lungs and digestive organs destroyed by what they had been breathing and eating. An official of the Kuwait Oil Company said publicly that one couldn't properly give the name "air" to the mixture of gases and oily particles circulating between the desert surface and the smoky shroud that rose from the oil fires. Later, when Kuwait's government had reestablished itself in the country, the company would claim, incredibly, that the air in Kuwait City was never polluted and that the fallout from the fires would fertilize the desert. In the chaos following the triumph of Operation Desert Storm, however, the evidence of one's smarting eyes and stinging lungs prevailed.

Around the middle of the month, the first scientific assessments of the catastrophe got under way. A group of Kuwaiti scientists who had been active in the resistance formed the Kuwait Environmental Action Team. They went into the field and took what measurements they could, but their laboratories had been demolished by the Iraqis, so it was impossible for them to analyze accurately the air and soot they collected. A team from the British Meteorological Office arrived with a laboratory installed in an aircraft and found that the smoke from the fires was generally densest at about 1½ miles and thinned out to practically nothing above 2½ miles. That information calmed some of the fears that the smoke would rise to more than six miles, above the cloud-bearing troposphere, where rain can flush the pollution back to earth. Any soot that got past the troposphere was fated to circulate for months, if not years, threatening to shade the whole earth and bring about a "nuclear autumn."

A task force of scientists from the United States, representing the Centers for Disease Control, the Environmental Protection Agency, and other federal organizations, began work in Kuwait. Under orders that appeared to have come down from President George Bush, the team, known as the "Interagency Task Force," could issue only innocuous statements about the war's environmental consequences—reporting, for example, that the "catastrophic predictions of the fires' effects are exaggerated." It was almost a month before they released their first findings, which emphasized surprisingly low values for the toxic gases hydrogen sulfide and sulfur dioxide.

By the beginning of April, it began to look as if the effort to cap the blazing wells would never itself catch fire, and the cry to expand the effort and introduce innovative methods was increasingly heard. The Union of Concerned Scientists, based in Cambridge, Massachusetts, convened a symposium to discuss different notions of how to win the battle against what NASA scientist Joel S. Levine called "the most intense burning source, probably, in the history of the world." The participants recognized that unexploded mines and other ordnance posed the most serious obstacle. They suggested that the military use helicopters to drag sleds of specially designed harrows across the mine fields or employ a jet of highly compressed air to clear paths through them. (The latter idea was actually tried in Kuwait before the fires were put out, although not in mine-clearing operations.) One symposium participant urged the use of explosive collars—which did not yet exist—around damaged wellheads to pinch well casings shut, or nearly so. Another suggested that funnels be placed over wells that were spewing oil but not burning; the funnels could direct the flow of oil into temporary storage pits, minimizing the extent of land contaminated by spilled oil.

Of course, all the ideas depended on ample supplies of either materials with which to fabricate large objects or heavy equipment to move large objects through the seething swamp that Kuwait's oil fields were becoming, and the shortage of such materials and equipment was exactly what had slowed the fire fighters' initial progress. Even had supplies been sufficient, the logistical problems of feeding and housing additional personnel in Kuwait would have remained. Told of the symposium's wealth of ideas for how he should modify his procedures, Red Adair asked if there was a single person at the symposium who had ever put out an oil-well fire.

On April 11, the Gulf Pollution Task Force of the U.S. Senate's Environment and Public Works Committee opened the first of four hearings. Brent Blackwelder, vice president for policy of Friends of the Earth, an environmentalist organization, testified that the initial report by the Interagency Task Force, which had been sent to Kuwait from the United States, contained conflicting information. Signed by William K. Reilly, administrator of the U.S. Environmental Protection Agency (EPA), it acknowledged the need for both increased environmental monitoring and warnings to Kuwait's residents who might be especially susceptible to the irritants that were burning the throats of virtually everyone. But according to Blackwelder, the report went on to emphasize the low values for dangerous gases and to note that "some data obtained by the team were encouraging." When the chairman of the committee, Senator Joseph Lieberman of Connecticut, asked Jim Makris, director of the EPA's Chemical Emergency Preparedness and Prevention Office, whether the air pollution in Kuwait was adversely affecting the health of Kuwaiti residents, Makris admitted that there had been a large number of respiratory problems in Kuwait, but added the view of Kuwait's minister of public health, who said that respiratory problems always increased in the spring because of airborne pollen and dust from the desert. Plant life in Kuwait did not produce much pollen in 1991, however.

At the same time, things were slowly improving in Ahmadi. Around the first of April, Russian Antonov cargo planes, the largest in the world, joined the fleet of carriers bringing heavy equipment, spare parts, and food for the fire fighters and support crews. After a few weeks' experience in Kuwait, the fire fighters were now able to develop better strategies for approaching and fighting the fires, improvising techniques that made efficient use of the limited supplies available. From Ahmadi, the crews fanned out and slowly began to kill the wells, one every few days. First they went after the wells that occasionally closed Kuwait International Airport with their smoke and mist. Then they put out blazing gushers that fouled the air in Ahmadi. Typically, road-building crews in tin-armored bulldozers pushed causeways of explosive-free desert sand from the main roads toward the flaming wells. Stretching across the ponds of burning or smoldering oil, the causeways were thick enough to absorb the impact of any exploding mines buried along their route.

Just below the sand that covers most of Kuwait is a layer of gypsumlike clay, known locally as gatch, which the crews used to top their new roads. Under a steady load of traffic, it hardened into a pavement capable of supporting bulldozers and the like. For a day or two, each new gatch road, with its almost-white surface, stood in stark contrast to the black landscape that it crossed; then the inexorable shower of soot and petroleum mist brought it into conformity with its grim surroundings.

Bechtel Corporation, the principal contractor for most of the goods and services required by the fire fighters, began in April to deploy crews that dug pits for storing water. Each pit had the capacity to hold between 500,000 and 1,000,000 gallons of water. The expatriate workers on Bechtel's crews eventually dug more than 200 pits and lined them with heavy-duty plastic. By April 4, they had completed their first three pits and converted an underground gas pipeline into an aqueduct that carried sea water from Mina Ahmadi, on the Gulf coast, to the pits in the Ahmadi field, the northern section of the Greater Burgan field.

A fourth fire-fighting company, Safety Boss of Calgary, Alberta, arrived on April 7, bringing with it several 50,000-gallon tanker trucks. Safety Boss's president, Mike Miller, figured that the ponderousness of the Houston-based companies' operations was at least partly responsible for the modest number of wells capped—fewer than thirty on the day his crews arrived—so he founded his operation on the concept of highly mobile units that could be rapidly set up and dismantled at a location. Not needing enormous amounts of water or gargantuan machinery, Miller and his crews were able to arrive at a blowout, organize their water and dry-chemical supplies and pumps, put out the fire, and move off to another blowout, often in less than a day. They used cranes to position a length of wide-diameter casing pipe, usually about ten yards long, over a blazing gusher and then ran hoses carrying water and dry chemical into the bottom of the pipe. The smaller fires usually succumbed after only a few minutes. Capping a blowout after extinguishing the fire was grimy and dangerous work—the oil could reignite—but capping crews from all the fire-fighting companies usually accomplished the job without casualties in less than a day after a fire had been put out.

Boot & Coots discovered a different way to make do before Bechtel established regular water supplies. From their long years of experience, Hansen and Matthews knew that there were plentiful supplies of liquid nitrogen in Saudi Arabia, so they had King arrange to have tanker trucks of it brought in. Using the liquid nitrogen in much the way the Safety Boss crews used dry chemical, Boots & Coots crews pumped the minus-320-degree material into the bottom of a well-casing chimney and succeeded in smothering a well fire on April 7.

By April 12, thirty-two wells, four of which had been on fire, had been brought under control. The liquid nitrogen technique soon became unnecessary, as Bechtel was steadily improving the availability of water. By April 26, crews from Safety Boss, Red Adair, Boots & Coots, and Wild Well Control had capped sixty wells, including twenty blazing ones. King felt as if his logistics operation was beginning to rival that of the Gulf War itself. In Ahmadi, people hoped that, if they didn't have to drill any relief wells, they might be able to wrap the job up in eight to ten months.

Things didn't look so bright, however, for Kuwait's minister of oil, Rashid al-Amiri. Within the emir's *diwaniya,* or council of closest advisers, al-Amiri was seen as largely responsible for hampering the delivery of supplies and equipment. Although the pace of operations was slowly increasing, all too often the lack of a spare part as trivial as a small spring could render ten hours' work on a blazing well useless. Al-Amiri tried defiantly to hold on to his position, acknowledging repeatedly that the effort was progressing too slowly but suggesting that it was the fire fighters who were to blame and vowing to bring in more crews from other companies around the world. When, on April 22, the emir replaced him with Hamoud al-Roqba, all agreed that the change pointed the way toward a more efficient operation and better relations between the Kuwait Oil Company and the fire fighters.

But before the end of the month, five people, in two or possibly three separate incidents, burned to death after being engulfed in the flames of an oil lake. None of the fire fighters was surprised; indeed, they had been grimly wondering when the first fatality would occur. A few days earlier, a crew from Wild Well Control had extinguished a wellhead fire, and members of the crew had walked up to the still boiling-hot wellhead. After inspecting it to their satisfaction and slogging back to their trucks, they heard a tremendous whoosh, turned, and saw the oil gushing from the wellhead spontaneously erupt into a demonic blowtorch once again. If it had reignited a minute sooner, they would all have been consumed in its flames.

The first incident, on April 25, occurred when the natural-resources editor of the *Financial Times*, David Thomas, and photographer Alan Harper drove their car along a road in the Burgan field south of Ahmadi, a location where officials of the Kuwait Oil Company had advised them not to go. Their car apparently skidded into an oil lake that had nearly flooded the road. As they splashed into it, the heat from the car's catalytic converter apparently ignited the dense petroleum fumes around them. They had no chance to escape; the air they gasped for was itself on fire. A Safety Boss fire fighter drove past shortly afterward and noticed the burnt-out remains of the car but assumed it was one of the hundreds of wrecks that had littered Kuwait's landscape since the end of the war. Later in the day, a pumper truck carrying two Indian employees of the French oil-services company Dowell-Schlumberger slid into the same lake, which reignited, killing them both. To complete the disaster, a tanker truck driven by an Indian employee of the Saudi company Hussein Nassr, which had been following the pumper truck, slid into the same deadly trap.

CHAPTER TWO

Success in the Oil Fields

"WHAT THIS OIL'S
gonna do to this country five or ten years from now, it's
terrible to think of, but for us right now it's a hell of a lot
more important that we somehow get to the wells so we can
kill 'em," T. B. O'Brien said in the slow drawl common to
the men who have made their living in the oil business.
O'Brien coordinated the relationship between the Kuwait Oil
Company and the Houston-based fire-fighting companies.
In early April he traveled south from Ahmadi with Sa'ud al-
Nashmi, overall director of the well-control effort for the
Kuwait Oil Company, to the emir's farm, a man-made walled
oasis in the desert, planted with palm and tamarisk trees and
housing its own menagerie. In April 1991, thick flows of
burning oil swirled through the once pale buildings, now
stained black by soot and petroleum fog. The palm trees,
fronds burned away, looked like telephone poles painted with

creosote. The air stank, and it stung their noses, while the only light under the blackened midday sky came from the fireballs of the nearest burning wells, exploding again and again as they rose into the gloom.

O'Brien's mind was occupied envisioning a strategy for getting long-reach backhoes, bulldozers, and hydraulic cranes to the burning wells, and securing supplies of high explosive to blast away damaged wellheads and blow out the fires. He speculated on how Bechtel was progressing with its plans to deliver millions of gallons of water a day to the desert, and how Paul King's helicopter missions were succeeding in locating heavy equipment that had been abandoned in the desert. When he turned to al-Nashmi, he saw tears running down the Kuwaiti's cheeks. "I've worked my whole life to build up this country and its oil industry. Look at it now. We'll never have it back the way it was," al-Nashmi cried.

❦ ❦ ❦

AROUND THE MIDDLE of May, the attack began on the well known as Ahmadi 120—so called because it was the 120th well drilled in the Ahmadi field. Even to Raymond Henry, who had overseen the effort at the Piper Alpha disaster in the North Sea, where 165 people were killed on the night of July 6, 1988, the wellhead looked unapproachable. One gigantic fireball after another was erupting vertically, two blowtorches were roaring out horizontally at ground level, and the whole well was encircled by a fiery swamp of oil-drenched sand. Like a number of other wells blown up by the Iraqis, Ahmadi 120 had had its casing shattered several meters below the desert surface, and oil was escaping both from the casing and from the three howling ruptures. Oil oozed up through the sand to the surface, where it ignited. Gushing about 40,000 barrels a day, Ahmadi 120 spewed enough oil an hour to have powered fourteen average American cars for 100,000 miles had it been gasoline.

To gain access to the well, Henry had Bechtel's bulldozer drivers build a ramp of sand across the outer moat of
fire and dig a trench to drain away unburned oil from the
well. Oil continued to ooze up through the ramp and reignite, but the bulldozers repeatedly piled the sand and packed
it down, and after a few days the sand was compacted sol-
idly enough to prevent any oil from moving through it.

Once the fire fighters were able to approach the well,
they brought in the long-reach backhoes and dug down to a
point below the lowest casing rupture. Because the roar that
a blown-out well makes would give pause to the ground crew
at any major airport, and because the heavy-equipment op-
erators sit with a shield of corrugated tin between their backs
and the fire, Henry directed the Indian backhoe operator
with hand signals, sometimes relayed through Brian Krause,
a well-control specialist. They dug down thirty feet before
they came to clean casing, all the time trying to stay upwind
and keep the drain clear so the gushing oil wouldn't swamp
the hole. It was a bizarre conjunction of twentieth-century
technology and medieval fantasy—knights on foot or on
massive steeds in combat against a gargantuan dragon that
heaved out sulfurous gusts of flame.

Gathering the entire flow into one vertical gusher meant
removing all the damaged parts of the wellhead. The quick-
est way to get rid of a wellhead as far gone as Ahmadi 120's
is to blast it with several hundred pounds of plastic explosive, a dirty method since the blast is sure to cause its own,
not completely predictable, damage. But the relatively controlled damage caused by a deliberately set charge is always
more amenable to repair than the mess that the crew then
faced.

On the third day of the effort, Henry guided a crane
operator toward the well. He always had the operator take
at least one practice run up to the well so that the men
operating the water cannons could set their pumps to the

correct rate and aim the water jets precisely. The operator also benefited from knowing all the adjustments he would have to make in maneuvering the explosive mounted at the end of the crane's boom up to the wellhead. Henry usually stationed his water cannons about 65 feet from the fire, inside tin huts. Without the tin, the heat radiating from Ahmadi 120 would have raised blisters on the men's flesh within minutes. One spray showered the tip of the crane's 65-foot boom at a rate of some fifty gallons of water a second, to keep the explosive cool. After the crane operator had placed the explosive precisely where Henry had intended, everyone at the site ran for cover, and Henry detonated the charge. Ahmadi 120 was more than equal to the blast. The three fissures continued to belch flame, which coalesced into a thick column of black smoke.

The next tool Henry and Krause tried was a half-inch steel cable, which they managed to wrap around the well below the lowest fissure. They attached its ends to winches set back about 150 feet from the well; one winch would pull in a few dozen feet of cable, then the other would pull it back, and so forth, in an attempt to saw off the wellhead. Crew members sprayed the cable with water to keep it from melting. Krause figured that it would take a few days, possibly even a week, to saw through the wellhead, but once the job was accomplished, he would have a clean cut to work with when he capped the well. In the meantime, nobody at the site could tell whether any progress was being made because oil and sticky sand were constantly blowing into the hole around the well. On the fourth day of sawing, the cable broke.

Back in Ahmadi that evening, Henry, Krause, and the crew discussed what to do next. They speculated that the sawing might have notched the wellhead deep enough for a bulldozer to snap it off, a trick that had worked in the past. The next day Henry ducked under the three mouths of the wellhead, threw a 1½-inch cable around its neck, and hitched the cable to one of the bulldozers. The force of the 'dozer turned out to be less than the resistance of Ahmadi 120 but considerably more than the strength of the cable. It broke with a snap that could be heard even above the roar of the flames.

When the crew returned to Ahmadi 120 on the ninth day of the job, Henry planned to continue whatever progress the cable saw had made. But he could not find among all the twisted and scarred metal on the wellhead an obvious new groove. So he wrapped a new length of cable around it as best he could and once more set the winches to work. The machines yanked the cable back and forth for six more days before it finally gave out, but Ahmadi 120 still stood. The crew tried pulling it down with the bulldozer and thick cable again, but that cable also broke. After three more days of sawing, there was another broken cable.

At that point, Henry decided to use plastic explosive, but he had no better luck than he had had on the first attempt eighteen days before. He and Krause had decided to employ the explosive in the form of a "shaped charge"— plastic explosive packed into a cavity cut out of a slab of metal. When the explosive detonated, the strength of the metal would direct the force of the explosion in a single direction, out from the cavity. By the next morning, the metal shop in Ahmadi had prepared a 10- by 14- by 24-inch piece of metal with a wide V-notch cut into one of the 10- by 24-inch sides. At the location of Ahmadi 120, Henry packed about seventy-five pounds of explosive into the notch, secured the charge to the end of the boom on the crane, directed the crane operator toward the well, and detonated the charge. When he looked to see whether this stratagem had worked, there was Ahmadi 120, still howling and blazing as hellishly as ever.

It was clear to Henry and Krause that they would have

to do something new. They named the product of their new idea, in the overworked parlance of the Gulf War, "the mother of all shaped charges"—300 pounds of explosive nestled into a cone-shaped cavity cut into one end of a steel cylinder 28 inches long and 30 inches in diameter, "by far the largest shaped charge either of us had ever seen or heard of," said T. B. O'Brien.

Once more Henry directed the crane driver up to the well, again the fifty gallons of water a second kept the explosive cool against the blazing 1,000-gallon-a-minute gusher, and once more everyone hunkered down and ducked. All, that is, except a Kuwaiti general who happened by and couldn't resist the temptation to witness the spectacle. The force of the explosion blew out the fires as though they were birthday-cake candles, obliterated the top of Ahmadi 120, and pushed over what was left of the wellhead at a thirty-degree angle. Killing the well would be a routine task now, and could wait until the next day.

The general praised Henry, Krause, and the rest of the crew for their skill and determination, and O'Brien thanked the general, suggesting at the same time that the crew would be very grateful if he could procure one of the jet cutters—devices that use a high-pressure stream of water and sand to saw through the bolts that hold the flanges onto the Christmas tree—that were said to be available.

Within a couple of days, the general fulfilled the request.

⬱ ⬱ ⬱

ALTHOUGH THERE WAS certainly rivalry among the first four companies operating in Kuwait, they had agreed during planning sessions in October 1990 to set competition aside and cooperate as much as possible. Each company brought its own supply of pumps, tools, and Athey wagons—small cranes mounted on bladeless bulldozers encased in heat-protective tin—but in a pinch, if a crew from one company found itself in need of some piece of equipment or advice, it could count on the other companies' crews operating in the vicinity to provide what help they could.

Red Adair's company was easily the best known of the first four outfits. Press coverage of explosive oil-field disasters routinely included photos of his puffy but leathery face beneath a broad-brimmed tin hat, peering intently into yet another inferno. Adair's method was systematically to remove all obstacles before beginning to attack a wellhead. Once his Athey wagons had dragged away whatever detritus cluttered a location, he, Henry, or another senior fire fighter would inspect the wellhead at close range under the protection of a deluge from the several water cannons. The inspection usually revealed the secret to killing the well. Sometimes a relatively simple adjustment, such as tightening a bolt, can choke the fire-breathing dragon. Other times one might have to cut off the bolts of a damaged flange and replace it. On yet other occasions one can "sting into" the well with a device that, like a gigantic bee stinger, could inject drilling mud down into the face of a gushing well. The stinger connects through hose and pipe to a pump that shoots the mud through the device. If the pump is strong enough, and the drilling mud heavy enough, the burden of the mud will eventually hold back the force of the erupting oil. Altogether, Adair's crews capped 111 wells in Kuwait.

Boots Hansen and Coots Matthews brought considerably fewer men and less equipment than the other three main companies. But with only three three-man crews and four twenty-foot cargo containers of equipment and spare parts, they killed 128 wells.

Joe Bowden's Wild Well Control, founded in 1975, was the youngest of the Houston outfits. Bowden, who was on the first flight to Kuwait after its liberation, stayed for nine days before returning to Houston and gathering his forces. When he came back in late March, he brought six

4,000-gallon-a-minute water pumps, six pipe racks, and six Athey wagons. It was several weeks, however, before he could use the diesel-powered water pumps because the first pits that Bechtel dug were not near any of the wells his crews proposed to kill. Since he couldn't count on a continuous water supply, Bowden depended on water brought to burning wells by tanker truck. He and his men even killed seventeen wells in April by stinging into them without the benefit of cooling showers of water, and finished with a total of 137 to their credit, about 120 of which had been on fire.

Safety Boss took longer to sign its contract with the Kuwait Oil Company and did not arrive until the end of the first week of April. It had the largest crews, typically eight men, four of whom were blowout specialists. Its contract called for the Kuwait Oil Company to buy the sort of firefighting equipment that the Houston-based companies owned, including three trucks modified to a Safety Boss design. Safety Boss fitted these so-called Smokey units with high-capacity water pumps and fire extinguishers capable of delivering 200 pounds of dry chemical per second. Although Mike Miller, the company's president, credits the mobility of the Smokey units for his company's record of 180 wells killed, fire fighters from the other companies, in what sounded like a case of post-crisis sour grapes, attributed Safety Boss's success to a presumed focus on the smallest and most easily extinguished fires. Miller did not deny that his company killed more small blowouts than the Houston outfits but refuted the notion that the smaller fires were a particular focus of his crews. He said, "Each of four original companies was given sectors in which to operate, and we put out all the fires in our sectors just like the other companies put out the fires in theirs." Safety Boss was the only company operating in the Burgan field for several months, and it was the only company among the four principal ones that was hired for further work in Kuwait after all the fires were out.

By the beginning of May, the fire fighters began to realize that they were going to have the job finished in less than a year. They had capped sixty-eight wells in April, for a grand total of seventy-four (twenty of which had been aflame), or a tenth of the exploded wells. But it was slow going for the first few months. Oil lakes, especially in the Burgan field, were constantly moving into new areas because of changing winds and the unrelenting addition of millions of gallons a day. Meanwhile, researchers at the U.S. National Oceanic and Atmospheric Administration's meteorological observatory, perched at 11,000 feet near the top of Mauna Loa in Hawaii, discovered that soot from the fires had doubled the normal concentration of particulate matter in the air. Journalists reported continued complaints by the fire fighters that Kuwaiti officials were delaying the delivery of equipment. And around many wells, deposits of coke, a kind of petrified ash, began to accumulate. Some developed into volcano-like cones more than ten feet high that the fire fighters had to tear away before they could inspect the well itself.

Wherever the oil lakes spread, so did ground fires. Around virtually every blazing well, oil-soaked soil repeatedly ignited until the soil was hot enough to boil away the water that the fire fighters sprayed on it. Even after they would put out the well fires, the ground was often so hot that ground fires would spontaneously combust and reignite the blowouts. After a while, the fire fighters wouldn't shut off their water cannons until they had the well capped and the ground cooled down.

Still, the eight crews operating in Kuwait during early May were killing increasingly difficult well fires at a rate of almost two a day, and provisions of water and equipment were steadily improving. Kuwait's new oil minister, Hamoud al-Roqba, appeared willing to address the fire fighters' needs more quickly than his predecessor had been. No drilling derricks hampered access to the wells, and despite flow

rates up to twenty-five gallons a second from 4,800 feet below the surface, a high percentage of the wells came under
the classification of "low-pressure." And many of the wells
weren't damaged as severely as they might have been.

This meant that, once a crew put a fire out, it was a fairly
routine procedure to sting into the well, without having to
spend the time—sometimes days—to lop off a damaged
wellhead, have a crane bring a new blowout preventer, and
bolt it on. The supply of jet cutters improved, too. With
high-pressure blasts of sand and water, they could cut through
the bolts of a damaged flange in a tenth of the time that a
hydraulic bolt cutter required. The fires from low-pressure
wells could be put out with water alone, and although a fair
number of wells, like Ahmadi 120, took weeks to kill, they
were in the minority.

The fire fighters struggled to cap fifty-two wells in May
and only forty-eight in June, but in July the number rose to
sixty-eight, for a grand total of 245. As word spread that the
crews from Houston and Calgary were handling most of the
wells with ease, more companies from the United States joined
the effort, helping to boost the capping rate to better than
four a day during the first part of August. On the ninth of
that month, all the exploded wells in the Ahmadi field had
been brought under control. On September 8, about twenty-
four weeks after the first well was capped, the Kuwait Oil
Company announced that half of the exploded wells had been
killed.

While the fire-fighting effort was on, Kuwait was scrambling to ready its export terminals and refineries so that the
restored wells could start generating revenue as soon as possible. The first tanker to leave Kuwait with a load of crude
oil departed in the middle of June, carrying almost 800,000
barrels (with possibly 8,000,000 barrels a day still gushing
from the wells). By the end of July, repairs at Mina Ahmadi's export terminal, the site of one of the war's most

devastating spills, had progressed far enough that the first loaded tanker in a year left from there. Another milestone was passed in late July, when Kuwait's controlled production of crude oil reached 140,000 barrels a day, an amount equivalent to the country's consumption of refined petroleum products. But the amount of oil still gushing dwarfed controlled production until well into October. At the beginning of November, when the capping was declared accomplished, Kuwait was producing about 320,000 barrels of crude oil a day.

September and October had seen the arrival of fire-fighting crews from all over the world, hired by the Kuwait Oil Company, some people believe, as a political gesture of thanks to the nations that took part in the liberation of the emirate. A Chinese crew, which capped ten wells, began work on September 13. Kuwait fielded its own crew, which began operations on the fourteenth and eventually bagged forty-one wells. A Hungarian team that entered the scene on the twenty-eighth and capped nine wells brought two of the most bizarre fire-fighting contraptions anyone had ever seen: modified Russian tanks with jet engines mounted on top. Blasting a mixture of water and fire-extinguishing chemical, they made for wonderful footage on the 6 o'clock news but failed to arouse much enthusiasm among the Texans, who observed that most of the time the jet engines just blew the fire over and, in fact, caused ground fires.

Joe Bowden claimed the additional crews slowed the operation down, but the crew from Romania, which arrived on October 7, capped six wells; a French team, which began on the tenth, added nine wells; and a crew of Britons, which took the field on the sixteenth, capped six. The last team to arrive came from Russia. It started work on the twentieth, stayed for nine days, and capped four wells.

Through the summer and into the autumn, the increasing numbers of crews in the field capped the gushing blowtorches at a steadily climbing pace, reaching a record high of thirteen wells in one day. Each week, the restoration of Kuwait's roads and other transportation facilities made it easier to provision the crews. As soon as Kuwait's bureaucrats returned to their posts from the countries where they had sat out the occupation, they began processing new work permits for truck drivers, welders and cooks, who were eager to return from their native countries in southern Asia to the high-paying jobs available to them in Kuwait. Nearly 10,000 workers from thirty-four countries participated in the work.

Paul King, the American charged with providing all the equipment, oversaw the provisioning of kitchens, which prepared more than 3,500,000 meals (and up to 26,000 in a single day) for the fire fighters, their supporting crews, and the office and domestic personnel. He had more than twelve tons of ice imported per day. As the ordnance-clearing teams completed their work in the Gulf sea lanes and along the road from Saudi Arabia, destroying about 20,000 mines and other explosives, ships and trucks were able to deliver an increasingly diverse menu.

By November King had amassed about 6,500 pieces of heavy equipment, including items scavenged from the desert during the early days of the effort—the largest nonmilitary vehicle and equipment fleet in the world. Ships and trucks had delivered more than 140,000 tons of it, and supercargo planes like the Antonov and C5A, another 5,500 tons. At the height of the deliveries, as many as eight ships and six airplanes a day were arriving with cargoes that Kuwait will continue to use as it rebuilds its devastated oil fields, pipelines, and refineries.

The equipment fanned out from Ahmadi into the blazing oil fields along 175 miles of new roads. To build the roads, 2.4 million cubic yards of gatch had to be dug up—enough to build a pyramid 100 yards wide and 540 yards high. Bechtel also dug 361 lagoons in the oil fields, each with a capacity of between a half million and a million gallons. Water supplies came from the Gulf through about 250 miles of pipeline and hose capable of delivering at as high a rate as 25,000,000 gallons a day, or less than a tenth the average rate at which oil was gushing out of the ground.

Oil Minister al-Roqba placed the value of the lost oil at $12 billion, and the cost of the fire-fighting job at $1.5 billion, but others on the project said that the Kuwait Oil Company paid more than $2 billion to put out the fires. Figures on the total amount of oil lost are equally varied. The company claims that about 3 percent of the country's total reserves were lost, but some reports place the loss as high as 10 percent. If indeed "only" 3 percent spilled and burned, the total would be just under 3 billion barrels, which means that an average of nearly 11 million barrels gushed from the earth per day during the 265-day holocaust; that is a little less than twice the amount of motor gasoline that was being consumed per day in the United States during the Gulf War—6.7 million barrels.

In addition to the five men killed on April 25, there were two other casualties: one man stepped on a mine, the other was run over by a bulldozer. Serious burns sent five men to the hospital for extended treatment, and two others suffered relatively minor burns.

Kuwait celebrated the fire fighters' triumph in the oil fields on November 7, when, in the presence of foreign dignitaries and representatives of the fire-fighting companies, the Emir of Kuwait, His Highness Sheikh Jaber al-Ahmad al-Jaber al-Sabah strode across a red Oriental carpet spread out on the sands of the Burgan field and stepped up to a lectern draped in blue cloth. There he pushed a lever, closing a valve on a nearby well that had been capped a few days before and subsequently reignited for the ceremony. Kuwaiti men in traditional dress performed sword dances for the assembled group. To symbolize the success of the effort, a truck delivered a load of clean sand to cover a nearby area of oil-soaked soil. That evening, residents of Kuwait City cruised back and forth along the shoreside thoroughfare, Arabian Gulf Street, in their cars, honking their horns as they went. Even with the backdrop of the emir's ruined al-Seif palace, the scene was a happy one.

The festivities in the Burgan field were billed as the killing of the last well, but in fact that event was still in progress up north in the Sabriya field. The following day, a Safety Boss crew there succeeded in its battle against the final well fire, which was one of the most spectacular of them all. In the eight-plus months that the well had been gushing and burning out of control, the desert soil around it became saturated with oil, leading to possibly the largest and most dangerous ground fire that any crew faced. Enormous coke deposits that hindered access to the well were also the result of months of burning. To top it off, the flow out of this last well had been increasing as the other wells in the vicinity were capped. The contrast to the emir's ceremony of the previous day could hardly have been more sharply drawn, but the Safety Boss crew killed this last dragon just as surely.

With the killing of the final well, hundreds of millions of barrels of oil remained in the desert, thickening slowly. Some Kuwaitis considered that the effort to restore their country's environment had only just begun; the sad reality was that it was coming to a close.

CHAPTER THREE

The Biggest Oil Spill in History

≪ ≪ ≪ ≪ ≪ ≪ ≪ SOUTH OF KUWAIT, wait, along the Gulf coast of Saudi Arabia, the saddest sight was the birds. They died by the thousands in the oil, and their plight was the first image that television crews focused on once the oil drifted out of the military restricted zone. Confused by petroleum fumes, the cormorants, grebes, sea gulls, and wading birds mistook the deadly slicks for the salty waters of the Gulf, swooped in to land, and found themselves trapped, mired in a sticky goo that no amount of preening could remove. Not knowing what else to do, they continued to preen, eating the oil picked from their feathers until their bills were glued shut. Then, twisting their bodies and flapping their wings in a grotesque dance of death, they struggled to free themselves from the swamp that was slowly smothering them, drawing them down deeper with every kick and jerk. Finally, they had strength enough only to straighten their necks listlessly for a moment, futilely try to open their mouths and give a call of distress, then let their heads flop down into the acrid ooze again. You could see dozens of them in one sweep of the eye, some so far gone that they could manage only a little bob of the head every several minutes or so.

During most of 1991, the Gulf was transformed into a stinking sea of death, fouling its own shores, bed, and even the air above it. Up to eleven million barrels of oil entered the Gulf because of the war, and that deluge was just the beginning, rather than the conclusion, of the war's attacks on the Gulf. The war's legacy has stuck onto rocks and penetrated into sediments along a stretch equal to the distance between Boston, Massachusetts, and Portland, Maine. And there nearly all of it remains, assuming, like the oil lakes in Kuwait, its place in the geological record of the Gulf region.

You could smell it hundreds of yards away if you walked across the claylike desert soil known as *sabkha* that extends inland from the low dunes near the Gulf shore. When you reached the water, the enormity of the spill overwhelmed you. Turning the air greasy with its fumes, it stretched to the hazy horizon north, east, and south and fell heavily on the shore in viscous waves too burdened by their own weight to resemble a normal surf. It was no mere slick that slopped onto the desolate coast of northern Saudi Arabia; it was a tide of crude oil. Walk ten yards out into it and still there was no water below, just sand, sopping the oil up like a sponge. At sea, windblown sand from the desert drove the oil down to the bottom, where it despoiled the seabed and everything living there for miles out from shore. At high tide, it washed over thickets of stunted mangroves and lawns of the strange-looking plants called halophytes, coating leaves, stems, and roots with a choking layer of tar. Twice a day for weeks on end, the tides threw up the clinging goo and left a gargantuan streak across 400 miles of shoreline.

GULF COAST OF SAUDI ARABIA'S
EASTERN PROVINCE

One thing you didn't see much of on the beaches of Saudi Arabia's Gulf coast during those first weeks of the spill was people. With the exception of occasional military units on training maneuvers, the numerous peninsulas and dendritic inlets between Kuwait and Abu Ali island were deserted.

Sandy beaches form the shoreline for about 60 percent of the distance between the Saudi Arabia–Kuwait border and Abu Ali. It is a stressful environment for the ghost crabs and beach fleas, the most noticeable shoreline animals there. They have to cope with sand that, under the Arabian sun, can heat up to 160 degrees Fahrenheit and that is swept away and redistributed with every tide. The Gulf itself is a little larger than Kansas and is one of the world's saltiest seas. Its principal sources of fresh water are the combined output of the Tigris and Euphrates rivers and the scant three-quarters of an inch of rain that falls on it in an average year. In the summer, its surface waters can warm to 100 degrees and produce the hazy blanket of humidity over the Gulf. Salinities may exceed that of average seawater by 15 percent; in the shallow bays and lagoons along its western shores, salinities rise to 45 percent above average seawater.

Despite the austere beauty of the coast, virtually all seaside recreation takes place on the other side of the Arabian Peninsula, at the Red Sea. There is no coastal road in the northeast. To reach the Gulf shore by land you have to take the main road north from Dhahran, which passes through the Greater Burgan oil fields to Kuwait City. About 60 miles northwest of Dhahran, you leave the pavement and start driving across the *sabkha*, trying to stay on an established track with enough speed to avoid getting stuck in the spots where some moisture has turned the soil into quicksand. You see mirages of water in every direction, but the Gulf is to the east. During the first few months of 1991, you could depend on your nose to lead you there—once you cleared the innumerable military checkpoints, that is.

The spill drifted south from a war zone where independent sources of information were almost nonexistent, representing the worst nightmare of environmentalists everywhere—an ecological cataclysm unfolding on top of the world's fountainhead of petroleum. For weeks, one could only guess at the total quantity of oil issuing from sources whose number and locations were obscured by wartime confusion and secrecy. The military spoke at length about the discharge from Kuwait's Sea Island loading terminal, about ten miles offshore from Fahaheel; but there were also numerous reports of discharges from Iraq's offshore loading terminal and from tankers that had been targeted by Coalition air strikes. Whether by design or simply because there were too many other things to deal with, the military did little to clarify the most elementary questions about the oil spill, even for Saudi environmental officials. The first reports, which ultimately proved to be amazingly accurate, spoke of quantities in excess of 10 million barrels, stunning a public whose standard for an oil-spill disaster—the *Exxon Valdez*—was one-fortieth the size. Almost immediately, however, conflicting estimates asserted that the earlier reports were wildly exaggerated, and that the total spilled probably did not exceed a million barrels.

These questions about the spill persisted for about a month after Saudi Arabia's Meteorology and Environmental Protection Administration began coordinating an overall response in late January. Not until Coalition forces liberated Kuwait on February 27 did it become possible to obtain corroborating reports on the oil flowing from Kuwait and the northern Gulf. Even then, however, a gag order, which appeared to have originated in the White House, prohibited U.S.-sponsored scientists from making all but the most innocuous statements about the war's ecological consequences—even to their peers at conferences abroad. Government-imposed secrecy continued to cloud the issue until a month after Kuwait's liberation, when journalist John Horgan, writing in *Scientific American*, reported on the information embargo. On April 3, 1991, in an Op-Ed piece in *The New York Times*, Tom Wicker hailed Horgan's work and suggested that the U.S. information blackout was an element of a pattern based on the fear that ecological catastrophes would dampen enthusiasm for the war. In the meantime, however, Kuwaiti citizens and representatives of the United Nations had begun to assess the quantity of oil in Kuwait Oil Company storage tanks and to compare the figures with those recorded in company inventories before the invasion. Thus they were able to determine the amounts of oil released by Iraqi forces. They were able as well to locate not only pipelines that the Iraqis had rerouted to dump crude oil into trenches along the shoreline or directly into the Gulf but also a number of sunken tankers leaking oil.

Only in the summer were the facts revealed about the much-heralded U.S. bombing mission credited with stanching the flow from the Sea Island Terminal on January 26, 1991. According to a briefing by General Norman Schwarzkopf on the day following the mission, a successful attack by a U.S. Air Force plane armed with precision-guided missiles destroyed manifolds that controlled the flow of oil from storage tanks located about five miles inland from Mina Ahmadi to the Sea Island Terminal. The combination of that attack and an engagement between the U.S. Navy and a vessel suspected of laying mines near Sea Island—a skirmish that set fire to the oil flowing from the terminal—resulted, Schwarzkopf claimed, in reducing the spillage from Sea Island "to a trickle."

The bombing mission appears in fact to have had almost nothing to do with stemming the flow of oil from Sea Island, which reached a total of 6 million barrels before it finally petered out in the middle of March. And the spillage would probably have been considerably worse had not a handful of Kuwaitis hoodwinked Iraqi troops into thinking that another set of storage tanks was emptying its contents into the Gulf. What seems to have happened is that the bombing mission frightened the Iraqis into abandoning the facilities that controlled the flow of oil out to Sea Island. Taking advantage of the Iraqis' absence, Kuwait Oil Company employees entered the facilities and closed valves that had allowed oil to flow out from the storage tanks. The Kuwaitis then switched the "open" and "closed" indicators on the valves to fool the Iraqis on their return to the control facility.

The massive spill began almost immediately after the beginning of the air war, when the Iraqis dynamited Sea Island. At about the same time, they scuttled five tankers moored at Mina Ahmadi. It was apparently oil from those tankers, about 2.5 million barrels of it, that provided the war's first footage of spilled oil, a video taken before January 25 in the vicinity of al-Khafji, just south of the Kuwait–Saudi Arabia border. Despite numerous claims that a large discharge ran into the Gulf from storage tanks at al-Khafji, there is no evidence of such a spill: fighting in al-Khafji at the end of January led to the destruction of one large petroleum storage tank there, but the oil that was in the tank was consumed in a fire.

During February, Iraqi sabotage at the oil ports of Mina Ahmadi, Abu Halifa, about five miles to the north, and Shuaiba, immediately to the south, resulted in additional discharges. The Iraqis also dug long trenches along the coast, filled them with oil in expectation of an amphibious invasion by Coalition forces, and eventually emptied them into the Gulf. In addition, thousands of barrels of oil drained into the Gulf from the oil lakes that formed under the hundreds of gushing wells on land. The spillage into the Gulf was particularly acute from the Sabriya field, north of Kuwait Bay, where a natural drainage system of dry watercourses, or wadis, is developed to a greater extent than anywhere else in Kuwait. Oil continued to flow down the wadis until the bulldozer crews supporting the fire-fighting activity built up levees to hold it back.

With virtually every bit of information passing through military censors before appearing in print, on radio, or on television, some questioned whether Iraq was ultimately to blame for the catastrophe. In fact, Coalition attacks on Iraqi targets clearly played a role in two of the largest discharges. In the first of these incidents, on January 23, U.S. Navy fighter jets attacked the Iraqi tanker *Almutanabbi*, which was equipped with antiaircraft guns. The tanker was moored in the northern Gulf, providing cover for Iraqi Hovercraft gunboats. Perhaps Saddam Hussein calculated that the Coalition forces would be unwilling to attack the tanker and risk the embarrassment of causing a large spill. But war is war. The *Almutanabbi* disgorged more than 900,000 barrels of oil into the Gulf. The other large release resulting from Coalition military actions originated at Iraq's offshore loading terminal, Mina al-Bakr, a facility similar to Kuwait's Sea Island Terminal. The U.S. military acknowledged that Mina al-Bakr was a target of air strikes but claimed that Iraq caused the spill by repeating the actions it undertook at the Kuwaiti facility. In response to a reporter's question about what the U.S. military planned to do about the spill from Mina al-Bakr, General Schwarzkopf said that if the spill from the terminal went "out of control" the military would "do exactly what we did with the other one." About 700,000 barrels of oil entered the Gulf at Mina al-Bakr.

The only unassailable facts about the combined spillage during those last weeks in January were that it was enormous and that it gushed from the mists of the war zone. Personnel from Saudi Arabia's Meterology and Environmental Protection Administration (MEPA) risked Iraqi antiaircraft fire to conduct aerial reconnaissance missions aimed at determining the size and locations of the slicks that made up the spill, but the flights were turned back when they reached the restricted airspace over Kuwaiti territory. For the first three weeks after the initial reports, all the observers could tell, apart from the position of the slicks in Saudi waters, was that, from as far as they were allowed or they dared to fly, oil extended northward to the horizon. After the middle of February, smoke from the oil-well fires added to the wartime fog that obscured reliable information on the spill. When MEPA officials tried to verify press reports of spills, all the military would tell them was that the bulk of the oil was northeast of al-Khafji, that the situation was being monitored, and that they would be informed if the situation changed significantly.

Saudi Arabia had established a round-the-clock response center in Jiddah on January 21 but moved into high gear on the 25th, when MEPA requested the assistance of Captain Roger Mowery of the U.S. Coast Guard in organizing the response. Mowery, a liaison officer stationed in Bahrain at the time, immediately flew to Dhahran, met with Saudi officials, and initiated the entire U.S. response. Getting the right people into Dhahran as fast as possible was a touchy problem because of the war. MEPA personnel from Jiddah flew in by circuitous routes to avoid Scud attacks. On January 27, when MEPA's president, Abdulbar al-Gain, gave his first press conference in Dhahran on the spill, a Scud attack sent participants scurrying for their gas masks. The conference continued when they returned, masks ready.

Among the first spill-response experts to arrive in Saudi Arabia was David Usher, president of Detroit-based Marine Pollution Control. He had also been among the earliest experts on the scene after the tanker *Exxon Valdez* went aground on Bligh Reef in the Gulf of Alaska on March 24, 1989. Even though Prince William Sound is a more remote location than the Gulf coast of Saudi Arabia, volunteers and professionals quickly poured into the area, eager to begin cleaning up the nearly 11 million gallons of spilled crude oil. Usher's crews safely off-loaded about 40 million gallons of oil from the stricken tanker, preventing the spill from surpassing a million barrels—all while the tanker was in danger of breaking into pieces. Usher could hardly be more pointed in contrasting the Alaska and Persian Gulf spills:

In Saudi Arabia, we were coming into a war zone. Our departures from the United States and our arrivals in the Gulf region were kept secret for fear of terrorist attacks. I went over because the United States government asked me to go, but there was no way that I was going to ask any of my people to risk their lives fighting a spill that you couldn't even get near the source of. Besides, what kind of company would jump up and ship its equipment, without liability guarantees, into a situation where a Scud might land on it? We all knew that it was a big spill, and since the military wasn't about to give anyone any information that might affect its operations, all we had to go on for a month or so was that it was a big spill and we were going to have to deal with it where we could.

❦ ❦ ❦

EVEN IN PEACETIME, Saudi Arabia is a remarkably difficult destination. The country issues no tourist visas, so one must have business to attend to or be on the hajj—the Islamic pilgrimage to Mecca—before the government will permit one to enter. With a war in progress, it was an invitation-only affair. Iraq's Scud missiles were overhead; mines drifted in the Gulf waters. Wartime insurance premiums would have had to be paid for any emergency personnel and equipment going into Saudi Arabia to clean up the spill. Because the premiums were especially high for those venturing out into the Gulf, hopes were dim for collecting most of the oil at sea and preventing it from fouling the shore. But shore-based cleanup had its own obstacles. The claylike *sabkha* that forms much of the soil near the shoreline resembles quicksand, and it takes a skilled eye to negotiate a journey across it, even in a four-wheel-drive or tracked vehicle.

The quantity of oil involved in the Gulf was also of a different magnitude from other spills. The largest spill prior to the Gulf disaster, the blowout at the offshore well Ixtoc I in the Gulf of Mexico, took nine months during 1979 and 1980 to spew its 3.3 million barrels into the Bahía de Campeche before Red Adair finally capped it. The 1983 spill of 1.9 million barrels from Iran's Nowruz and Ardeshir oil fields in the northern Gulf also played itself out over the course of months. With the spills of the Gulf War, an enormous amount of oil entered the water almost instantaneously, and oil continued to spill into the Gulf at a rate of up to 6,000 barrels a day for as long as three months, until the second half of May. Seven sources were still leaking as of the latter part of June.

In 1991, researchers at Saudi Arabia's King Fahd University of Petroleum and Minerals in Dhahran used data from a Landsat satellite to estimate the size of the spill. Satellite images of the military restricted zone were not available to the researchers, however, until the cease-fire, and with the Gulf bristling with mines, they had no way of verifying their estimates, or getting the "ground truth," of the oil's thickness. Visual estimates taken from airplanes and helicopters flying over the Gulf ran up against similar limits. Even after the U.S. Coast Guard agreed to a Saudi request and sent over two jet aircraft equipped with the state-of-the-art oil-detection system known as Aireye, the best estimates still included guesses about the percentages of the oil that had either evaporated or been absorbed into sediments. The Coast Guard planes' important role in keeping track of the oil's position, however, allowed computer modelers at the university to predict where the slick would be in days to come, which in turn helped MEPA officials decide where their scarce resources would best be deployed.

The Aireye oil-detection system is a computer-controlled combination of side-looking radar, an infrared/ultraviolet scanner, and a reconnaissance camera. At altitudes of between 5,000 and 8,000 feet, the system gives good results out to about 40 miles on either side of the aircraft. The radar detects differences in wave heights, which the computer interprets as damping caused by an oil slick; the scanner detects differences in the reflection of sunlight off the surface, from which the computer calculates differences in temperature. The computer also provides output on oil thickness by using formulas that relate greater temperature differences and stronger reflections to increasing oil thicknesses. Finally, the computer uses information from the camera to pinpoint each area of an oil slick.

Early on, a number of environmentalists started complaining that the response to the disasters of the Gulf War was inadequate and was perhaps being hindered by forces within either the Saudi or the American governments. But the Saudis acted with characteristic deliberateness and all due seriousness. With the help of the United States and the United Nations International Maritime Organization, they quickly assembled the most competent group of experts available. MEPA planners knew that little could be done to prevent unprecedented quantities of oil from polluting their coast, so they focused their attention on ways, first, to collect as much oil as possible at the shore and, second, to minimize the "collateral damage" that would inevitably scar the Gulf and the coast in the event of a large-scale cleanup campaign. While environmentalists cited the fact that 11,000 people took part in the *Exxon Valdez* cleanup, the Saudis wondered if 11,000 people on their beaches would mitigate or exacerbate the effects of the spill.

"The oil spill was without precedent," recalled MEPA's Abdul-Jaleel al-Ashi, "so we immediately hired the best people in the world to help us. We very carefully evaluated the most beneficial techniques to use under the circumstances. We did not want to do something that might make the oil disappear today, only to cause problems one year, ten years, or fifty years later. The last thing we were about to do was to put thousands of people on our beaches with shovels in order to make a good impression on television screens around the world."

Still, environmentalists around the world clamored for Saudi Arabia to "do something" about the spills, for although the televised coverage of Operation Desert Storm included heart-wrenching footage of cormorants struggling in a sea of oil, the television cameras revealed no evidence of crews deploying boom to corral the tarry mess or skimmers to gobble it up. Indeed, the sorriest victims of the delayed effort to protect natural areas were the birds, despite a well-organized attempt to rescue and clean them. Hastily built on the outskirts of al-Jubail by Saudi Arabia's National Commission for Wildlife Conservation and Development, a rescue center began treating birds on February 8. Among the volunteers who flocked to the center from distant lands was John Walsh, a ruddy Australian who is director of the World Society for the Protection of Animals. The National Commission organized the center with the cooperation of the Royal Commission for Jubail and Yanbu. At the center, wildlife specialists from abroad joined with an enthusiastic contingent of expatriate engineers whose industries had slowed down because of the war and off-duty military personnel to rescue and rehabilitate as many of the afflicted birds as they could. Because the center was organized so quickly and so many foreigners were available as volunteers, it rapidly developed into virtually the only opportunity in the Islamic kingdom for members of the opposite sex to meet one another in a public setting. When word of the center's social benefit got around, the number of volunteers swelled until the kingdom's religious police insisted that if women wanted to volunteer they had to have a center of their own. The National Commission then organized the women-only center, and the number of volunteers leveled off.

Walsh and other volunteers trudged the oil-soaked beaches

to the north of al-Jubail and brought more than a thousand birds back to the centers by the end of March. Most of them were cormorants, either great or socotra, and grebes, black-necked or great crested. The socotra cormorants alone represented about a third of the birds rescued, and they were of particular concern because they are a threatened species that nests primarily along the shores of the Gulf. About 60 percent of the birds at the centers died, as did thousands more that were never brought there. "The birds we treated represented only a fraction of the total mortality," said a downcast Walsh.

One morning, as Walsh and a handful of volunteers searched along a 1¼-mile stretch of coral and sand, they picked up about fifty live birds and counted more than that dead. As Walsh looked back to the area they had covered, he saw more birds crawling out of the chocolatey ooze, which lapped onto the shore in layers several inches thick. He watched a narrow offshore coral reef on which many birds were standing, already covered in oil by the outgoing tide. As the tide came in, the birds struggled in the thick oil but could not swim. They slowly sank, with only their heads showing, then only their beaks, then nothing.

CHAPTER FOUR

"We Must Recover the Oil"

F LYING ABOVE THE Gulf coast of Saudi Arabia in early February 1991, Jim O'Brien, an oil-spill contractor from New Orleans who had helped manage cleanups at both Ixtoc I and the *Exxon Valdez*, gazed down on the curling fingers of the innumerable small inlets and bays between Ras al-Tanaqib and Ras al-Zawr and imagined strategies for corralling or absorbing the gelatinous monster below. Oil covered the water and the shoreline as far as the eye could see north and south. From the low-tide line, it obscured Gulf waters for more than two miles out. The land was almost as flat as the water, so flat in places that the distance between the high- and low-tide lines was over half a mile. An inches-thick coating of oil dripped heavily over every square foot of the intertidal zone, making instant fossils of birds, fish, crabs, and worms. O'Brien knew that all his techniques for herding spilled oil toward areas of minimal ecological sensitivity were already useless. There was no way to "finesse" this spill.

He couldn't depend on much help from nature, either. Saudi Arabia's Gulf coast is a classic example of what oceanographers call a low-energy shoreline. No crashing storm waves are ever going to scour it clean. Its slope is so gentle and subtle that every rising tide inexorably carried the heavy oil up to the high-tide mark, and as the tides ebbed, an even coat of the stuff was left over plants, animals, and sand. For all O'Brien—or anyone—knew at the time, the oil might continue flowing from Kuwait into the salt marshes for months.

O'Brien's flight over the Gulf coast came about two weeks after Saudi authorities received their first word of the spill. On January 24, a day before Saudi Arabia contacted Captain Roger Mowery of the U.S. Coast Guard, MEPA vice president Nizar Tawfiq had told Hosni Abdurasek, his administrator in Dhahran, that an oil spill of unknown proportions had occurred and that MEPA must amass all available expertise and equipment to deal with it. At that moment, MEPA was in virtually the same position as the first fire fighters to arrive in Kuwait would be. It did not have the means with which to do its job. The kingdom had put the agency in charge of coordinating damage control less than two months before. And now, armed only with the country's original contingency plan—which had been approved just a week before Saddam Hussein ordered the discharge lines opened at the Sea Island Terminal—without oil-spill experts, sufficient equipment, or any emergency funds, MEPA's officials had to come to grips with an unimaginable deluge.

Tawfiq and Abdurasek began by appealing for the assistance of qualified people everywhere. President George Bush announced on January 27 that he would send a nine-man team, which later included David Usher of Marine Pollution Control as well as representatives of the Coast Guard, the

Environmental Protection Agency and the National Oceanic and Atmospheric Administration. With experts from other countries, this task force came to be known as the International Interagency Assessment Team. By February 1, the U.N. International Maritime Organization had received offers of equipment and expertise from more than ten countries. As it was still more than a week before the Iraqis began to blow up Kuwait's oil wells, environmentally conscious people the world over fixed their attention on Saudi Arabia's Gulf coast.

Within a day of the task force's arrival, Tawfiq chaired the first general meeting on the crisis. In a conference room in MEPA's still unfinished Dhahran headquarters, he sat before a huge map of the Saudi coastline, from al-Khafji to Dhahran, and faced a crowd of MEPA personnel, officials of various Saudi organizations, foreign experts, and the media. The throng spilled out the doors and into the hall. With the outward calm that Saudis characteristically adopt when dealing with momentous responsibilities, Tawfiq defined the problem at hand and outlined the groups' duties. Every individual and every organization should plan on being taxed to the limits of their resources, he said.

The unfinished headquarters and a building at the Dhahran airport became centers of activity overnight. Hundreds of workers, dressed in traditional Saudi garb, in Western clothes, or in oil-splattered coveralls and boots, gathered every bit of pertinent information available, issued and carried out orders, pored over ambiguous reports, and reached decisions based as much on trust in Allah's benevolence as on anything else. Every day, early in the afternoon and late in the evening, Tawfiq presided over a general meeting, where heads of his departments and other agencies reported on the day's progress, the new challenges, and the persistent problems. At least once every day, Aziz al-Amri, a young MEPA employee, reported on that day's aerial missions to determine the positions of the slicks, the ropy water-oil emulsions known in the spill-cleanup trade as "chocolate mousse," and the thinner oil sheens that made up the overall spill.

For more than a month, al-Amri's reports were the only reliable accounts of the oil's position because no information was forthcoming from the military on activities within the war zone. Flying two and sometimes three missions a day, al-Amri often returned to Dhahran airport with bloodshot eyes brimming with tears from the heavy smoke of the oil fires, which rose to 13,000 feet, hampering his surveillance activities until the middle of July. Passengers in Saudi National Guard helicopters, making passes at 300 feet over the spill, usually got headaches from the smell of crude oil that hung in the air; by flight's end, their skin, the helicopter, and everything in it would be coated with a greasy film of petroleum that had evaporated from the slicks below. On the smokiest days, the soot and petroleum fog almost closed the airport, as the light from the sodium-vapor lamp standards penetrated only about 200 yards.

Since Saudi Aramco, the largest industrial concern that owned facilities threatened by the oil, had trained crews and quicker access to large stocks of equipment, its representatives cochaired the daily meetings with Tawfiq. Inevitably, the organizations with the greatest resources tended to assert power. Jealousies rose to the surface, and there were sharp criticisms on all sides. But citing the call in the Koran for Muslims to pull together toward a common goal, Tawfiq succeeded in getting the various industries and governmental bodies to act in the spirit of the contingency plan that MEPA was charged with coordinating.

The plan, however, was not exactly a how-to guide for cleaning up an 11-million-barrel oil spill. It had been conceived to deal with spills similar to those of the past, not the unprecedented disaster that served as its first test. The best contingency plans incorporate training programs based on experience gained from previous spills, and stress prevention first, then concentrate on controlling the spill source as quickly as possible, and finally come to grips with containing the spill as close to the source as possible. To have devised an effective plan in this case, MEPA would have had to have displayed extraordinary skill in international diplomacy and expertise in wartime countersabotage, and been able to assemble a flotilla of commando oil-skimming vessels to brave the mine-infested waters off Kuwait. Moreover, there was no ready supply of earth-moving or mobile oil-pumping equipment except what was owned by the municipalities and industries along the coast, which were, understandably, less interested in joining the general effort than in keeping their equipment on hand to protect their own investments: they were not about to jeopardize the safety of their seawater intakes and harbors for what appeared to be a hopeless battle to protect the ecological integrity of beaches that no one ever visited anyway. Summing up the situation before a group of reporters in Riyadh on the day after the U.S. Air Force made its raid on the pipeline manifolds near Mina Ahmadi, MEPA president Abdulbar al-Gain admitted in a press conference, "this spill represents a threat that was outside the scenarios considered in the recovery plans developed in the wake of the Nowruz spill."

Saudi Arabia's contingency plan placed the highest priority on protecting resources vital to human life, which meant seawater desalination plants. Three of these facilities lay along the Gulf coast of Saudi Arabia, supplying the kingdom with about 40 percent of its drinking water. Among them is the world's largest—a 220-million-gallon-a-day behemoth at al-Jubail—which provides Riyadh with 60 to 70 percent of its drinking water and serves the principal industrial area, a metropolis of oil refineries, petrochemical plants, and steel mills, nearly all of which use seawater for cooling purposes. Of the two other desalination plants, only the 64-million-gallon-a-day facility at al-Khobar was to be protected, since the third plant, at al-Khafji, which normally produced more than 3 million gallons a day, had been shut down since the outbreak of hostilities.

While the intake at the al-Jubail desalination plant was especially important, the military took their supplies of fuel chiefly from two Saudi refineries on the Gulf coast, so the refineries' seawater intakes had to be protected as well. One of the refineries was near al-Jubail, at Ras Tannura; it is one of the world's largest, with a capacity in excess of a half-million barrels a day. The other important refinery was to the north, at Saffaniya. The troops depended on Saffaniya to supply them with about 1.5 million barrels of fuel a day. Since it is north of Ras Tannura, and was therefore likely to be threatened by an oil spill from Kuwait sooner than the huge refinery to the south, forward-thinking Aramco officials had sent boom—a kind of long floating barrier to oil— and skimmers to Saffaniya from southern locations before the oil began to flow. Keeping the refineries in production meant more, however, than just supplying the military. Saudi Arabia was able to maintain its level of financial support for the war only because of its oil exports, which ran at a daily average of about 8 million barrels; this necessity led to the inclusion of tanker terminals in the highest-priority sites.

Within a day or two of the discovery of the spills, crews hired by the Saline Water Conservation Corporation (SWCC) had cordoned off the 360-yard-wide entrance to the intake basin at al-Jubail and encircled each of the plant's five intakes with an additional two lengths of boom. Although al-Gain announced that there was no risk to the kingdom's desalination capability, SWCC continued to refine its fortifications at al-Jubail, heeding a prediction of "heavy impacts" by February 10. Because Norwegians had helped to design the protection at al-Jubail during the Iran-Iraq War, SWCC asked Norway's State Pollution Control Authority for help. On the advice of the Norwegians, crews rearranged the boom across the entrance to the basin into a V configuration, with the vertex of the V extending out into the Gulf.

For large oil spills, teams usually deploy boom more to deflect floating oil into natural or artificial catchment areas than simply to hold oil back. Oil carried by strong currents easily slips beneath boom strung directly across its path. One of the principal concerns of a spill-response manager is to understand the natural forces that cannot be controlled and to deploy equipment so that it works with rather than against them. The Gulf's tidal currents are relatively weak, so wind is normally the driving force in surface flow. Consequently, variable winds created nightmares for the managers who tried to position boom so as to take advantage of wind-driven currents. Just when a large quantity of oil had collected in one area, a shifting wind would send it streaming in the opposite direction, away from collection crews and into the open water.

Fortunately, currents at the seawater intakes of Saudi Arabia's desalination plants and industrial facilities flow in one direction only, so the crews that arranged the boom at al-Jubail and other important intakes could be confident that it would deflect much floating oil to either side of the intake basins. At al-Jubail, they also arranged additional boom at angles to the sides to deflect away any oil that might slip under the first line of defense. In addition, they suspended cylinders of netting around the opening of each of the five intakes to catch suspended tar balls and mousse that had sunk below the surface. Jan Nerlan, one of the Norwegians, commented on February 15—when much of the Saudi shoreline from the northern border to Ras al-Zawr was already smeared with oil—that the missing piece in the protection scheme was sufficient equipment to skim off oil that might enter the basin. "We'll need more skimmers," Nerlan said, "if we get hit with a big raft of oil."

None of the organizations was satisfied with its supply

of tools. Everyone was scrambling to acquire skimmers, vacuum trucks, any kind of pump that could handle oil-water mixtures, and boom in all its various configurations: heavy-duty boom to corral large areas of heavy oil, lighter boom to string across entrances of critical water intakes or environmentally sensitive marshes, and absorbent boom to soak up sheens that might have slipped by. Once an assortment of boom was in place, crews constantly adjusted and fine-tuned their arrangements. As long as there was no clear limit to the amount of oil that might have hemorrhaged from Operation Desert Storm, and therefore no way to predict where or when the oil might wreak its most devastating effects, MEPA, Aramco, and other organizations in the Eastern Province of Saudi Arabia sought additional equipment and qualified expertise wherever it might be found.

Aramco had access to two of the largest stockpiles of spill-response equipment in the world, but still it went shopping for more. As a member of the Oil Spill Service Center in Southampton, England, it enlisted the aid of Britain's Royal Air Force to airlift a hundred tons of boom and skimmers to Saudi Arabia, the largest single shipment of supplies into the kingdom and about 25 percent of the center's stockpile. Aramco is also a member of the Gulf Area Oil Companies' Mutual Aid Organization, known by its virtually unpronounceable acronym GAOCMAO, reputedly the world's largest collective stockpile of spill-response equipment, which it distributes among its member companies in Saudi Arabia, Bahrain, Qatar, the United Arab Emirates, Oman, Iran, and Kuwait. GAOCMAO gave Aramco some assistance during the crisis, but since its other member companies were facing the same threat, it was compelled to spread its resources more thinly than it would have if the Gulf War spill had been better defined from the outset.

On the same day that television network CNN began running its footage of oiled cormorants on beaches near al-

Khafji, Saudi Aramco's U.S. subsidiary, Aramco Services Company, put out an all-points bulletin in an attempt to identify the stocks of boom and skimmers available in North America; by February 1, it had begun purchasing a good deal of the available equipment and airlifting it to Saudi Arabia. Within a week, boom and skimmers began secretly arriving at Saudi Aramco facilities under tight security for fear of terrorist attacks.

The International Maritime Organization (IMO) began coordinating most of the transfers of equipment to Saudi Arabia under the terms of a convention that ninety of its member countries adopted at the end of November 1990. Although none of the countries had yet ratified the convention, thirteen responded to the crisis in the Gulf almost immediately, sending equipment into Saudi Arabia on every available flight. A number of countries sent personnel to IMO's London headquarters, where they staffed the emergency coordination center on a seven-day-a-week basis, compiling inventories of available equipment and keeping track of both Saudi Arabia's requests and the position of oil in the Gulf. By February 20, sheen extended for tens of miles offshore, all the way from Kuwait to Abu Ali, and heavy concentrations of oil had rounded Ras al-Zawr and entered ad-Dafi Bay.

In the only effort to mount an overarching campaign of environmental restoration, IMO announced the creation of its Gulf Oil Pollution Disaster Fund on March 15, 1991. Dozens of scientific missions have traveled to the Gulf and Kuwait for observations and measurements and have recommended steps that would mitigate the havoc wrought on the physical environment, ecosystems, and public health; despite these recommendations and impassioned calls from environmental organizations such as Friends of the Earth, which issued a statement saying that "no one is getting the job done now," the IMO fund, which received donations of more than $6 million, turned out to be the single largest pool of money dedicated to the rehabilitation of the damage to the region's natural resources. Saudi Arabia estimates that a large-scale cleanup of its affected shoreline would cost nearly half a billion dollars, or less than one percent of the $60 billion spent by Coalition countries to prosecute the Gulf War. The cleanup of the *Exxon Valdez* disaster, a fortieth the size of the Gulf spills, cost $2.5 billion, or almost three hundred times the amount raised by the IMO.

The governments that contributed most generously to the IMO fund were the United Kingdom (£1 million), the European Community (1 million ECUs), Japan ($1.5 million U.S.), the Netherlands (1.25 million Dutch guilders), Canada ($250,000 Canadian), Switzerland (300,000 Swiss francs), and Denmark ($150,000 U.S.). Other countries, including Luxembourg, China, and Germany, provided in-kind contributions. The United States, which spent $7.4 billion to carry out the Gulf War, contributed neither money nor equipment.

For the first five or six weeks of the crisis, all the work to secure funds, equipment, and expertise was insufficient to provide protection for the industrial facilities south of Abu Ali and the undeveloped shoreline to the north. Most of the protection went to the south, and, as luck would have it, essentially all the oil remained to the north. Winds kept the heaviest concentrations confined to near-shore waters, a phenomenon that could not have been predicted. As a result, the desolate northern shores absorbed repeated inundations, and most of the oil had no chance to drift far into the Gulf, where it would have been able to move south, posing a graver threat to the Eastern Province's industrialized areas and the states of Bahrain, Qatar, and the United Arab Emirates.

Tawfiq traveled to Bahrain in late February to meet with representatives of the Regional Organization for the Protection of the Marine Environment and explore the possibility of mounting a region-wide response to the catastrophe. "Every barrel you help us recover now," he declared, "is one barrel less that you will have to worry about later." But although the assembled officials expressed sympathy for MEPA's failure to secure adequate funding and agreed to continue discussions, they declined to extend their countries' assistance beyond their borders. Heavy equipment was in short supply everywhere. In Qatar, the deployment of protective boom was delayed for more than a week because no trucks were available to haul it from storage sites. One delegate acknowledged later that MEPA and the other environmental agencies of the countries bordering the Gulf didn't have time to think beyond protecting important industrial facilities and to go into a wider sort of operation where environmentally sensitive areas might be protected. Until MEPA could come up with a sufficient budget, he said, the effort to protect natural areas would remain disorganized.

The Research Institute at King Fahd University of Petroleum and Minerals, in Dhahran, rapidly identified the most important natural areas, enabling MEPA to concentrate its efforts where they would do the most good. These were typically tidal wetlands with very shallow water, tortuous and branching inlets extending sometimes for miles inland, and numerous tidal pools. The intertidal zones typically consist of halophytes, or salt plants, in the highest and driest areas, and mangrove thickets and expanses of algal mat toward the low-tide line. Although ecologists value algal mat because it stands at the base of the coastal food chain and grows exceedingly slowly, it is not otherwise an endearing substance—except, possibly, to those enthralled by the Chinese shar-pei dog's gross amount of loose skin. It smells and feels the way one would expect something called "algal mat" to smell and feel; freshly moistened by the tide, it swells and buckles until it takes on the appearance of an expanse of rumpled skin. Just below the low-tide line, lush lawns of sea grass shelter hundreds of species of shellfish, worms, and the young of many kinds of fish, including a number of commercially important ones.

After the Research Institute predicted that most of the oil would continue to hug the shoreline, Aramco and some of the other industries to the south of Abu Ali transferred boom, vacuum trucks, and bulldozers to Saffaniya, Manifa Bay, and Musallamiya Bay. The Saudi government streamlined the process for granting visas to spill-response specialists like Jim O'Brien and David Usher and moved crucial equipment through customs much faster than usual. But despite all efforts, geography, weather, and the war seemed to conspire against a successful campaign to save the tidal flats.

The quicksandlike *sabkha* that seems to be everywhere along the northeastern coast of Saudi Arabia presented one of the many problems faced by the crews. Bulldozer drivers routinely got stuck and sometimes even overturned while pushing groins or sand spits out into the Gulf to trap oil or while excavating storage pits for collected oil. At one point, a road compactor overturned as a bulldozer struggled to extricate it from a *sabkha* trap. Practically no one had a working knowledge of the essentially uninhabited northern coastal area, and in the resulting confusion, crews bypassed some particularly sensitive areas and occasionally lost lengths of boom to outgoing tides. In one valiant attempt to retrieve some boom that was drifting out of Manifa Bay on the tide, an Indian laborer jumped into an empty boat and went after it. Although he too drifted out, he wasn't able to keep up with the boom, and since there were no other boats on hand with which to launch a rescue mission, he was in danger of being swept out into the Gulf. Luckily, a rescue party found him late that night, shivering in the bottom of his boat, which had found its way into one of the other convoluted arms of the bay.

Throughout February, the war was a tangible reality for

the workers trying to protect Saffaniya and Manifa Bay. Because of the danger of mines, no one dared come near the suspicious-looking objects that floated by or were stranded in the intertidal zone. Smoke from the oil-well fires filled the air, and slick after stinking slick washed up and overwhelmed strings of laboriously placed protective boom. Pounding artillery fire in the distance and screaming fighter jets and Scud missiles overhead so unnerved one contractor at Saffaniya that after a single day on the job he packed up and left the country without a word to anyone.

Although a crew sometimes succeeded in stringing boom across a tidal channel and thereby saved a salt marsh or mangrove thicket, there were always more channels available to the oil. And just when the oil seemed to have left Manifa or Musallamiya bays to join the mainstream to the south, the wind would change and send it back into now unprotected areas. The inundations finally covered every foot of shoreline from the border to Abu Ali—a distance of 400 miles. Under the circumstances, it made sense to forgo protection and concentrate on recovery.

Nearly all the recovery operations were on land, causing further damage, unfortunately, to the already devastated shoreline. It was almost impossible to recover oil in nearshore waters, first because of the mines, but also because the water was usually less than three feet deep, too shallow for oil skimmers to operate in. Out in the Gulf itself, however, a number of deep-water vessels collected huge amounts of oil. Notable among them were two converted oil tankers, the *al-Waasit* and the *al-Oyoun*, which a Norwegian company had leased to Aramco just as the war was getting under way. The two vessels collected more than 40,000 tons, or about 300,000 barrels, of oil on their sorties into areas of the Gulf that the military had assured them were free of mines. Collecting more than 10,000 barrels on some days, they skimmed the Gulf from the beginning of February to the end of April, at which point offshore waters were sufficiently clear of oil to make continued operations impractical.

The crews that worked in the northern bays developed close friendships. Months later, Saudis who had spent weeks in the oily swamps of Manifa and Musallamiya bays would remember in amazement how united in purpose they had all been. "There was nothing else in my mind," Abdul-Jaleel al-Ashi of MEPA recalled. "Day after day we woke up early, got dressed, worked harder than we ever thought we could until there was no more light to work by, came back to our rooms, took showers that never really got us clean, and fell asleep before we knew we were in bed. The cleaning people at the hotel could tell what room we were staying in just by looking at the towels." Religion played a part in the unity as well. "To cause pollution like this deliberately goes absolutely against Islam, which instructs us to use the gifts of this world in such a way as to provide their full benefit to future generations," al-Ashi explained. "How could we not join together in strength to deny Iraq's ruler the satisfaction of this act?"

The gravity of the situation created so much tension that if, in the midst of a struggle with a line of boom, someone lost his footing and splashed into the goop, everyone else laughed in relief. Minutes later, the unfortunate individual would be part of the audience for someone else's pratfall. At the twice-daily meetings where crews reported on the amount of oil they had collected, quantities in excess of 10,000 barrels were greeted with riotous cheers and whistles. But reports of frustration—mired bulldozers or gummed-up pumps that prevented the collection of more than a few hundred barrels—prompted loud but good-natured hoots and hisses. Al-Ashi recalled: "Of course we felt bad at the end of a frustrating day, but the shouts at the meetings didn't make us feel any worse. Everyone knew what it was like out there, and everyone knew that tomorrow could be their day to hear the booing."

By March 8, heavy concentrations of oil extended out from the shore for more than six miles along the coast south from Manifa Bay in a continuous blanket that reached the northern shore of Abu Ali. Musallamiya and ad-Dafi bays were awash in an inches-thick layer of weathered oil covering about 2,000 square miles. Although it was becoming obvious that the best chance for recovering significant quantities of oil was to mount a large operation in the small bay known as Khaleej Marduma, which formed the southernmost pocket of the Gulf between ad-Dafi Bay and Abu Ali, funds were nowhere to be found. MEPA spokespersons repeatedly answered the question of when something really big might get under way with a kind of mantra: "As soon as we get some funding to hire the contractors who are here with their equipment, ready to go to work."

The contractors were a couple of U.S.-based companies with experience in the Middle East—Jim O'Brien's Oil Pollution Service, and Martech USA Inc. of Anchorage—and a company based in the Netherlands, Tanker Cleaning Amsterdam. Each of these companies had stationed a few executives in Saudi Arabia during the war to keep in touch with MEPA and stay informed about the spill as best they could. But they were not about to commit their considerable resources and trained personnel to a project of unprecedented dimensions without the benefit of contracts, and the Saudis hesitated to commit their treasury to contracts until the war and all its uncertainties were resolved. It wasn't until April, and after Japan had made a multimillion-dollar donation to the Gulf Cooperation Council relief fund, that Saudia Arabia provided MEPA with sufficient funds to hire the big companies.

For about a week at the end of February, MEPA's lack

of money resulted in Aramco's assuming operational control of the shoreline cleanups in the northern bays; after all, the oil company was the only entity in the country that had the ready cash to bankroll the efforts and therefore—theoretically—to coordinate them best. But as soon as Aramco began to pour money into shoreline relief, the potential sources of further funding inside and outside the Saudi government suddenly found other, more pressing needs for their resources. Aramco saw that it had singled itself out as the national sponsor of shoreline cleanup, a role it had never aspired to, and quickly withdrew its money. As a result of Aramco's halting step, the struggle became, for a time, more disorganized than ever.

Meanwhile, David Usher was busy in Dhahran, sifting through hundreds of assistance offers and sales pitches made to the Saudis and separating the useful from the inappropriate or misguided. "What a lot of people don't realize," Usher later said, "is that, once you've got a huge oil slick on the water, the newer technologies don't offer you much in the way of getting rid of the mess. Big oil spills are dirty, so the equipment you use to clean them up has got to work even when it's slopped all over with oil, tangled up in seaweed, and plastered with saltwater and sand." The Saudis respected Usher's experience and subscribed, as well, to the idea, increasingly popular among ecologists, that huge spills do most of their damage in a short amount of time and that technologies used to remove oil from sand or other sediments can often be as harmful to the environment as the oil is. So they weighed very carefully the offers of products that sounded too good to be true.

❦ ❦ ❦

NOT ALL THE offers were exactly altruistic. During the past ten years, especially since the hot-water rock-washing crews made the news during the *Exxon Valdez* cleanup, ecologically aware entrepreneurs have been trying to enter the market for equipment and supplies to clean up oil spills. Now, with a record-breaking spill lapping up along the Saudi shoreline, manufacturers and peddlers of all manner of cleanup paraphernalia descended on the Saudis. There were pitches for dispersants, which break slicks up into smaller and smaller pieces and eventually allow the oil particles to dissolve in the water. There were also "bioremediation agents," which are fertilizers or mixtures of fertilizers and bacteria; according to the bioremediation pitchmen, these products accelerate the natural phenomenon of biodegradation, converting the spilled oil into plant and animal tissue. And then there was the man with a "sinking agent," a substance more commonly known as sand. This product would adhere to the slick and, if delivered in sufficient quantity, drive the pollutant to the bottom. Perhaps he didn't realize that sandstorms are common around the Gulf, but MEPA realized that the natural sinking agents in the region were working well enough by themselves.

Saudis take pride in the way they consider possible solutions to a problem in a logical and open-minded way, so they invited many purveyors of spill-cleanup goods to their kingdom and gave each a fair chance to make his case. The Saudis also take pride in their tradition of hospitality; they treat every person who enters the kingdom as a guest, with all the attendant rights and responsibilities. Since Saudi Arabia is not a tourist destination, practically all the foreigners who came to help and possibly to profit from the spill had never been in the country before, and sometimes their misconceptions about Saudi society collided disastrously with their mercantile aspirations. Many a Western salesman, for example, was blinded by preconceptions of "third world" countries and supposed that Saudis would be all too eager to embrace the latest product of modern technology. He would often be unprepared for an MEPA official's cordial but earnest questions about the product. If, say, a chemical dispersant made oil "disappear," where exactly did the oil go? If indeed it remained in the water, although "disappeared," what was the scientific evidence that it would not harm the Gulf in the future? Wouldn't it be likely to have an ecological effect, possibly a bad one, in the future? In Saudi Arabia, the custom is to weigh all possibilities graciously but exceedingly carefully; as a result, cocksure or indignant salespersons unprepared for the rigorous scrutiny and cautiousness often found themselves on the next flight home, their passages hospitably paid for by the Saudis themselves.

≼ ≼ ≼

DISPERSANTS WERE IN fact ruled out early on, though they are sometimes helpful in fighting spills. When the oil began streaming from Kuwait, Aramco wanted to use them— subject to MEPA's approval of the precise location, time, formulation, and method of application—and proceeded to perform trial applications. It discovered that the oil had weathered to the point where dispersants were ineffective, however, so mechanical methods were the only means used to get the oil out of the Gulf.

They work best against fresh oil spills in an open environment where strong waves can mix the chemical into the oil. In Prince William Sound, for instance, the U.S. Coast Guard preapproved their use because the Sound is a high-energy environment, and currents continually flush its waters out to the southwest along the Kenai and Alaska peninsulas and into the North Pacific Ocean. (As things transpired in March 1989, however, they were used in only a few tests.) The Persian Gulf, by contrast, is an almost entirely enclosed sea, sandwiched between the Arabian Peninsula and Iran, and open to the Indian Ocean only at the narrow Straits of Hormuz. A complete flushing of Gulf waters takes many years. The extremely constricted access to the ocean allows organic material to accumulate in the Gulf and enhances the likelihood that spilled oil, whether it has been artificially dispersed or not, will remain for a long time.

MEPA also shunned offers of dispersants because most desalination-plant engineers view dispersed oil as even a greater threat than an oil slick. A slick usually floats at or near the surface and can therefore be blocked with containment boom or collected with skimmers and absorbent materials. But since dispersed oil is scattered in tiny droplets from the surface to the seabed, it cannot be blocked by physical barriers.

Saudi Arabia's largest desalination plants use a process called multistage flash distillation to create potable water from the sea. It repeatedly boils and condenses incoming seawater in a series of chambers, with the condensate from each successive chamber becoming progressively purer. If chemical dispersants had been used on the Persian Gulf slicks, according to the engineers, dispersed oil could have easily entered the desalination plants. Once inside a plant's evaporation chambers, much of the oil would distill along with the seawater, instead of remaining behind with the salt, and wind up as a constituent of the plant's "fresh" water output.

But not everyone sees dispersed oil as such a threat to desalination plants. As refinery engineers know, the repeated boilings and condensations characteristic of multistage flash distillation produce a final product with a highly specific boiling point. Substances that boil at lower temperatures never get the chance to cool off enough to condense out, and so they pass straight through the plant as vapor. Substances that boil at higher temperatures never make it past the first few evaporation chambers, where they would cause a maintenance problem by gumming things up, although a certain amount of tarry buildup there would not affect the quality of the water flowing out of the plant. The only constituents of crude oil that could have significantly tainted the output of Saudi Arabia's main desalination plants

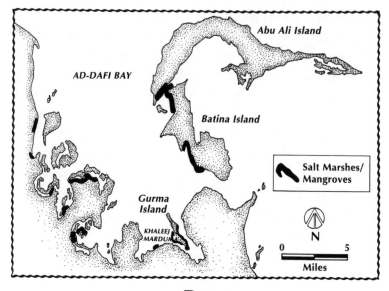

AD-DAFI BAY

were those chemicals that boil at precisely the same temperature as water, and even these, the engineers say, could have
been subsequently removed by various sorts of filtration.

No one, however, wanted an impromptu test of these
competing theories. All of ad-Dafi Bay and the waters north
of Abu Ali were coated by a rubbery blanket of oil eight
inches thick in some areas, and a fantastic opportunity to
recover an enormous quantity of oil off the water was developing there. The crescent-shaped island had become a gigantic natural boom and sponge, preventing the oil from
reaching crucial desalination and petrochemical plants immediately to the south. At the critical moment, the Royal
Commission for Jubail and Yanbu stepped in. A quasi-
governmental body that administers in Saudi Arabia's two
large industrial centers—al-Jubail on the Gulf and Yanbu on
the Red Sea—it had the political power to enlist the resources of the industrial concerns operating in the area. It
also essentially owned all the public-works vehicles in the

area and so had much of the equipment on hand to carry out a large effort; it was further able to summon the scientific support of the Research Institute at King Fahd University of Petroleum and Minerals. Through Aramco, the commission hired Seattle's Martech, and through the Saline Water Conservation Committee it hired O'Brien's Oil Pollution Service, or OOPS. Acting as agent for the commission, MEPA brought Tanker Cleaning Amsterdam into the operation. The Research Institute's forty-member group of computer modelers, using a spill-trajectory program called *Gulfslick 2*, determined that the greatest amounts of floating oil could be recovered over the longest period of time if operations centered on Khaleej Marduma and a small island in that bay, Gurma Island. "We must recover all the oil floating on the Gulf," said the commission's chairman, Abdullah ibn-Faisal ibn-Turki al-Saud, a Saudi prince. "We must be successful at Khaleej Marduma."

The prince's insistent declarations that "we must" use every available resource for the project resulted in the transfer of additional boom, electrical power generators, and heavy equipment from industrial locations in the south. Crews even redeployed boom to Khaleej Marduma from Tarut Bay, not far to the south, where Saudi Arabia's largest mangrove swamps are found. Saleh al-Obayyed, an official of the Royal Commission, was deeply moved by the cooperation that led to the additional equipment's transfer to the little bay, but he observed that "bringing all this material into play at Khaleej Marduma made our work all the more crucial; we had to be successful there, or the oil would have drifted toward the critical facilities where we had weakened our defenses."

Road-building equipment arrived to clear access to the sites on either side of Gurma Island, where *Gulfslick 2* predicted winds and tidal currents would steadily drive the oil. The Research Institute ran chemical analyses, such as flash point, on the oil to ensure the safety of recovery operations.

It investigated the suitability of various sites for excavating interim and long-term storage pits—a particularly important task since the Eastern Province has large aquifers that could be tainted by improperly stored oil. In addition, it used satellite data from NASA to develop precision weather forecasts and maps. Finally, the Royal Commission provided apartments, offices, maintenance facilities, clothing, food and water, communication systems, and hospital services for all the workers.

Jim O'Brien viewed the situation as if he were facing a monstrous mass that would smother anything in its path: "You've got to imagine it like there's this huge mess on the move; it's looking for a place to go, and it's got pressure. If you don't relieve that pressure, it's going to overwhelm no matter what you put out there." The problems required him to think in unconventional ways. "We had to get the oil, not necessarily into secure storage, but at least out of the way so that more oil could get to where we were ready to soak it up," O'Brien recalled. "When you have a massive spill like this, I think you've got more of an engineering problem in some senses than you do anything else. You have to think almost on a scale like road construction, improvise, and use bigger pieces of equipment to relieve the pressure of the oil that's bearing down on the shoreline."

O'Brien fought the battle with the help of hard-working Saudis and the Royal Commission's regular staff of a hundred or so Filipinos, Pakistanis, and Sri Lankans. They started recovering impressive amounts of oil on about March 10, operating shallow-draft sand dredges, called mud cats, that they had converted into skimming platforms for use in near-shore waters. The dredges' heavy-duty pumps were more capable than conventional pumps of lifting out the weathered Kuwaiti crude, and because the oil was coming in so thick, the dredges' intakes were able to take it up without bringing too much water along with it. From the dredges, large hoses carried the oil to holding pits running parallel to the shore, which became the hallmark of recovery operations. On shore, crews under O'Brien's direction dug eight trapping pits and built rough sluice gates into them. O'Brien recalled that "at high tide, we would open the gates and let the oil roll in, and when the tides went out we'd close them and hold the oil back." Like the oil sucked up by the dredges, the oil trapped in these pits was pumped to nearby holding pits, from which it was transported to dump-site cells lined with plastic. The plastic usually ripped, however, and underlying sediments got oiled anyway.

In meetings of the Royal Commission, MEPA, and other groups that played important roles in the Khaleej Marduma campaign, everyone contributed ideas about how to modify various pieces of heavy equipment to make them appropriate to the effort. Pumps were in short supply, so ways were devised to fabricate them out of stray parts, using the power takeoffs from John Deere farm tractors to propel them. With the prince's constant reminder that "we must be successful" in mind, O'Brien, on seeing one of the contraptions, said, "If that baby works, its name is the 'wemust' pump." Happily, the wemust pump served its purpose admirably.

Tanker Cleaning Amsterdam (TCA) built eight dikes out into the channel around Gurma Island and along the northern shore of Abu Ali and extended the dikes with booms that gathered the oil into the angle between the dike and the shoreline. Then it used its shallow-draft skimmers to pump the oil into holding pits, where the oil and water separated. OOPS, TCA, and Martech carried out their operations after brief discussions with the Royal Commission, which, according to TCA's Rob Lippens, "ran things with the idea that everybody has a considered point of view, and once all the points of view were on the table, they would elect a course of action, and that's the course of action we would take." All three contractors cooperated to their mutual satisfaction; when one group had skimmers or pumps to share, they unhesitatingly donated them to their "rivals."

One of TCA's contributions to the effort was quick-loading stations for the tank trucks that carried recovered oil to the dump-site cells. The trucks parked under raised tanks, which contained the oil pumped from the holding pits. A simple pull of a lever, and the oil whooshed into the waiting tanker. Although the procedure was quick, on occasion it proved to be dirty as well. Once, the outlet hose from the overhead tank came loose, and a few thousand gallons of thick, salty crude spilled over the truck and everyone within shouting distance. Another attempt to speed up the recovery—round-the-clock shifts—failed because nighttime temperatures cooled the oil to the point where it became thick and unpumpable.

Much of the work was carried out during Ramadan, which in 1991 began with the new moon in March. Despite the demanding physical labor, good Muslim workers kept the fast as faithfully as ever. One energetic MEPA worker, Abdul Halim, fervently fasted as he sloshed away in the oily muck of Manifa Bay, losing twenty-two pounds before the next new moon.

Even as some of the crews under the direction of O'Brien and Lippens were tearing up sandy stretches of shoreline for dewatering pits or collection trenches, other crews began cleaning the mangrove thicket on Gurma Island. Over the course of five or six weeks, they repeatedly flushed the island with sea water, slowly rinsing the heaviest deposits of oil from mangroves and halophytes, as well as from ecologically important thick carpets of algae. Lippens hoped that the flushings would accelerate the marsh's recovery from the 63,000 barrels of oil that he estimated had drifted onto the narrow, three-mile-long island.

The effort at Khaleej Marduma was the most successful oil-recovery operation ever. The contractors collected nearly a million barrels. During two months of work, a force that ranged between 150 and 250 men recovered an average of about 20,000 barrels of oil a day. The highest total for one day, recovered on March 14, 1991, was more than 67,000 barrels. Estimates of the cost for the Khaleej Marduma project range between $50 million and $100 million, a fraction of the amount the Exxon Corporation paid in 1989 to recover the much smaller spill in Prince William Sound.

Months later, in the New Orleans offices of OOPS, O'Brien reflected on how pleasantly streamlined the command structure in Saudi Arabia was in comparison to that in Alaska:

> In Saudi, we didn't have as many regulatory agencies to interface with, just a handful is all. But up in Alaska, we had the Coast Guard, Exxon, all the Alaska state agencies, the fishermen's groups, the local governments, and all of that. At least people got the chance to get their points of view across, but we were going to ten meetings to look at every problem. Over in Saudi, it was usually just one, and you're out getting the job done.
>
> It was good, too, that we didn't have to discuss the same things over and over again, because in terms of the amount of oil involved, the effort needed to fight the Gulf spills dwarfs the effort that was necessary in fighting the *Exxon Valdez*, the Ixtoc I blowout, or the *Amoco Cadiz*, a 1.6-million-barrel spill in 1978. If anyone tells you that they've worked on a spill cleanup of this scale, they're lying to you, because there hasn't been one.

In strictly economic terms, the effort at Khaleej Marduma will not pay for itself as long as the price of crude oil hovers at about $20 a gallon. But dozens of companies have offered to buy the salty crude, and Aramco, the official owner of the recovered oil, will probably recoup some of its expenses through such sales. Concerns over the possibility of being a third party to an environmental-crime lawsuit have led Aramco to study the proposed sales carefully. The Saudi company has already blended about 100,000 gallons of the recovered oil into its refinery stock, but the results of this experiment show that the salt in the oil could corrode a refinery's pipes and tanks. In its current storage sites—well-constructed pits surrounded by ten-foot-high chain-link fencing—the oil will remain isolated from the environment until 1994, according to MEPA.

Like Paul King, coordinator of the fire-fighting effort, Jim O'Brien was not surprised that no one reacted to Saddam Hussein's environmental hostage-taking by amassing all the equipment that would have been necessary for an immediate response. "Nobody would have had the desire to maintain a stockpile of that size, either," he said. "They certainly had a sizable amount of equipment over there before the spill occurred, much more than you would probably find, say, if you looked in U.S. ports. And the guys in charge of the equipment were well-trained and knowledgeable in deployment methodology and had even developed strategies about how they would operate in various situations. But to get all geared up ahead of time for the biggest spill in history just because Saddam Hussein said he would drown us in oil? You can't program for a madman."

CHAPTER FIVE

Dire Predictions,
Surprising Measurements

≋ ≋ ≋ ≋ ≋ ≋ ≋ SOON AFTER SADDAM
Hussein swore to destroy Kuwait's oil industry rather than
relinquish it, people all over the world began to speculate on
the effects his actions might have on the global environment.
The image that immediately came to mind was the chill of a
nuclear winter, and with it the fear of how far the smoke
might spread. If the invisible chemicals that escape from re-
frigerators and air conditioners can cause massive destruction of the earth's protective layer of ozone fifteen to thirty
miles above the planet, posing a threat to elements of the
planet's food chain, what incalculable damage might a catastrophic release of smoke and noxious gases from oil wells
bring?

The possibility of nuclear winter was a concept first discussed in 1982 as the likely outcome of a massive nuclear
exchange between the United States and the Soviet Union.

Enormous quantities of black smoke would rise high into the atmosphere, absorbing sunlight and shading the earth below, plunging much of the planet into a prolonged deep freeze. For this to occur, however, three elements would be necessary: many millions of tons of material would have to be injected into the atmosphere; the material would have to rise above the clouds, which would otherwise return it to earth in rain and snowfall; and the material would have to be black enough to serve as a planetary parasol.

History had recorded several oil-field conflagrations—a fact that was little discussed in the months leading up to February 1991. The most significant of them was small by comparison with the disaster in Kuwait, and it was not well studied at the time it occurred, but it was a precedent nonetheless. During the late nineteenth and early twentieth centuries, Azerbaijan was among the world's most important oil centers. In the 1880s, nearly two hundred refineries operated in Black Town, a suburb of Baku, which occupies a large peninsula jutting into the Caspian Sea at the southeastern end of the Caucasus mountains. Baku is the site where ancient Zoroastrians worshiped naturally occurring gas flares as "eternal pillars of fire." It is also the largest city in a region that has for centuries been the scene of vicious strife between Christian Armenians and Muslim Azeris—violence that erupted again even as the oil wells blazed in Kuwait.

During the Russian Revolution of 1905, Baku's Azeris rose up and attacked the wealthy Armenian upper class, which included the leaders of the oil industry. The Azeris set fire to every Armenian-owned property they came across in the course of their pillage. Eventually the fires spread to the oil fields, where wells had been drilled much closer together than is customary today. The resulting inferno was so terrifying that the survivors later compared it to what they imagined the last days of Pompeii to have been like. Smoke completely obscured the sun, enveloping Baku and its environs in a continual darkness that lasted for weeks, much like Kuwait eighty-six years later. The effects of the Baku fires did not, however, extend for thousands of miles.

So both a theory and a historical precedent pointed to the likely results of massive oil fires in Kuwait. No one doubted that the environment would suffer horribly when Saddam Hussein carried out his threat. The question was how far the devastation would spread.

Although the Kuwait Oil Company and Aramco started planning for environmental disasters ahead of time, they were concerned with a limited geographic area and driven by economic interests more than by a desire to save the world from an ecological cataclysm. They kept their discussions private, and it was not until King Hussein of Jordan gave a speech on November 6, 1990, at the Second World Climate Conference in Geneva, Switzerland, that the environmental implications of the impending war became a topic of public discussion. King Hussein stated baldly that the confrontation between Iraq and the Coalition forces led by the United States, impelled as it was by a "combination of human fear and political ferocity," and "taking place on top of the richest petroleum reservoir in the world," could lead "to an environmental catastrophe the likes of which the world has not experienced since the accident at the Chernobyl nuclear power plant."

He went on to to say that calculations by Jordanian scientists indicated that if Kuwait's oil reserves went up in flames, the environmental effects would spread far beyond the borders of the conflict, with smoke blackening the skies over Kuwait, Bahrain, Qatar, the United Arab Emirates, the Persian Gulf, and most of Saudia Arabia, Jordan, Syria, and Iran. The likelihood that war would lead to environmental ruin throughout and beyond the Middle East was so strong, King Hussein argued, that every attempt must be made to resolve the crisis through peaceful negotiations.

The speech was not well received. By refusing to join the U.S.-led juggernaut moving toward war, King Hussein had already incurred the anger of the United States and its Coalition partners. Prime Minister Margaret Thatcher of the United Kingdom brushed aside his environmental concerns, claiming that not only were they unfounded, they were also a dishonest front for the king's presumed desire to see Saddam Hussein establish rule over Kuwait. Saddam Hussein, on his part, repeatedly rebuffed the king's personal communications, in which the king cited numerous reasons for peacefully retreating from Kuwait before the Coalition forces went into action, as he was convinced they would if the Iraqis did not withdraw.

The preliminary calculations that King Hussein relied on in his predictions came principally from his science adviser, Abdullah Toukan, a nuclear physicist who received his doctorate from MIT. Shortly after Saddam Hussein announced his threat, King Hussein asked Toukan to determine what the implications of massive oil-well fires in Kuwait might be. Taking into account the basic unpredictability of the weather and of meteorological phenomena generally, Toukan assumed that if 500 oil wells began burning out of control, three million barrels of crude oil and its dissolved natural gas would go up in smoke each day for as long as the wells continued to burn. By limiting his prognostications to the effect of burning oil and ignoring the possibility of oil gushing to the surface and not burning, he underestimated the rate of spillage by up to 70 percent.

Prior to the invasion, the Kuwait Oil Company had been producing just under two million barrels a day from 864 wells, 363 of them in Kuwait proper and 501 in a zone where Kuwait and Saudi Arabia share oil production. By far the most productive were the ones in Kuwait, which accounted for an average 1.59 million barrels a day in 1989. They represented, however, less than half the 743 oil wells in the country. The other 380 were idle but ready to produce at any time. Tou-kan's assumption of 500 wells gushing and burning oil at a rate of 3 million barrels a day appeared reasonable in part because only 6 percent of the 363 producing wells in Kuwait were classified as "artificial-lift" wells, wells that require pumps to bring the oil up to the surface. In the rest, subter-ranean pressure drives it up from the depths.

Actually, the distinction between "artificial" and "natu-ral" lift is not so simple. Oil companies, including the Ku-wait Oil Company, routinely inject natural gas and other fluids into oil-bearing rocks that already produce by natural lift. The injected material maintains subterranean pressures and prevents the oil-bearing formations from falling in on themselves as oil flows to the surface. Ironically, Kuwait purchased its supplies of natural gas for this purpose from Iraq.

The pressure that propelled Kuwaiti oil to the surface was capable of ejecting far more than the country's OPEC quotas. Therefore the Kuwait Oil Company traditionally left most of its wells capped and choked the flow of producing wells in Kuwait proper back to a daily average of only 4,380 barrels per well in 1989. Toukan did not hazard a guess about the precise effect that the uncontrolled flow of oil from 500 oil wells might have on subterranean pressures, simplifying his calculations with the assumption that uncontrolled flow would be about 1½ times the rate of normal production. He thereby arrived at the even figure of 3 million barrels a day, a figure he estimated would hold for an indefinite number of months.

He predicted that the fires' emissions of soot and hazard-ous gases would present the greatest environmental danger. Toukan used *soot,* a term without a precise scientific defini-tion, to describe the particles of unburned organic matter given off in a fire. Its environmental threat lies in its black-ness, or ability to absorb sunshine. The blacker the soot, the more dangerous it is, because blacker soot shades the earth more effectively than less black soot does. Injected in sufficient quantities into the air high above the earth, its absorption of sunlight lowers temperatures and reduces plant growth. The danger posed by soot from oil-well fires increases with the height to which it rises. For if it rises above the level where clouds form, rain and snow falling from the clouds cannot cleanse it from the air. Although cleansing of the troposphere, the cloud-bearing layer that stretches from the surface up to about 42,500 feet in the Gulf region, aggravates the discomfort of those living near the fires, subjecting them to cooler temperatures and black rain and snow, it benefits the global ecosystem by limiting the time the pollution is aloft and consequently confining the pollution to smaller areas.

Gases certain to be released by the fires included carbon dioxide, the chemical most responsible for the greenhouse effect. Rains do not cleanse carbon dioxide from the atmosphere—only plants can, by incorporating it into their living tissues. Rains do, however, rinse the air of some ecologically significant gases given off by the fires, such as sulfur dioxide and the various oxides of nitrogen. Unfortunately, this cleansing comes in the form of acid rain, since both sulfur dioxide and nitrogen oxides are transformed on contact with water droplets into sulfuric and nitric acids, the principal agents of acid rain.

Toukan believed that the combustion in the oil-well fires would be incomplete, resulting in a situation much like the plumes of black smoke billowing from a diesel engine in need of a tune-up. More efficient, or complete, combustion produces less soot but more "greenhouse gases," especially carbon dioxide, than inefficient or less complete combustion does. By assuming that the fires would burn with relative inefficiency, converting 10 percent of the oil's mass into soot,

Toukan's calculations showed that, on the average, every barrel of oil burned would put 15.1 kilograms of soot, 108 kilograms of carbon dioxide, 14.3 kilograms of carbon monoxide, 3.4 kilograms of sulfur dioxide, and 0.85 kilograms of nitrogen oxides into the air. In a year, at 3 million barrels a day, the totals would be 16.5 million metric tons of soot, 118 million metric tons of carbon dioxide, 15.7 million metric tons of carbon monoxide, 3.72 million metric tons of sulfur dioxide, and 0.93 million metric tons of nitrogen oxides.

To have a global impact on the scale of nuclear winter, almost all the soot Toukan predicted would have to be injected into altitudes above the troposphere, where it would shade the earth from the sun's rays for a long time. Current scientific thinking is that about 105 million metric tons of soot would have to enter the atmosphere over the course of several days to engender a nuclear winter. About 6 percent of the crude oil burning as a result of nuclear holocaust would be converted into soot, according to this same thinking. Although such a global catastrophe was not in the offing, the quantity of soot that Toukan predicted would enter the atmosphere would nonetheless have devastating and possibly far-reaching effects.

Toukan realized that his theoretical soot needed an extra boost in order to reach the top of the troposphere, since his calculations showed that the heat of the fires would carry it only to altitudes of between 11,500 and 16,400 feet—not quite halfway to the upper boundary of the troposphere. So he raised the possibility that when sunshine hit the dark soot particles, the particles would warm up again and subsequently heat the air around them. Hot air rises, so the air that had been heated by the sun-warmed soot would carry some of that soot to yet-higher altitudes in a process known as "self-lofting." Toukan thought that self-lofting soot particles could reach the upper boundary of the troposphere within ten days after combustion and within several weeks be adrift throughout the Northern Hemisphere.

Mindful that the prevailing winds blowing across Kuwait come from the northwest and that Kuwait's yearly average of less than half an inch of rain provided little hope of washing the soot from the air quickly, Toukan suggested that the soot—even if confined to the troposphere—might disrupt normal weather patterns in Asia, specifically the monsoon rains. Subsequent research in a national laboratory of the United States has shown that his suggestion was prescient. Toukan did not speculate as to precisely the effect the soot might have on the duration or character of the monsoons, but he noted that southern Asia is the most densely populated region on earth and that the hundreds of millions of people living there depend on the regularity of the monsoons for their survival.

There was no doubt, Toukan concluded, that massive oil-well fires would devastate the environment near Kuwait. But the risk that the fires posed to the rest of the world was so grave, and so ill-defined, that he called for extensive computer modeling to pin down the soot's wider threat—a threat that could put much of the Northern Hemisphere at risk, a threat that the United States took very seriously.

On January 11, 1991, nine eminent scientists joined Toukan in a message to U.N. Secretary-General Javier Pérez de Cuéllar, stating that if the impending war resulted in the widespread destruction of oil wells, within thirty days soot could cover an area equivalent to half the United States, and the carbon dioxide released by the fires would add to the greenhouse effect. The group—which included Joseph Farman, discoverer of the "hole" in the ozone layer over Antarctica; Paul J. Crutzen, principal author of the original paper on nuclear-winter theory; and Carl Sagan, director of planetary studies at Cornell University—thanked Pérez de Cuéllar for attempting an eleventh-hour mission to forestall the onset of war and expressed hope that its findings would result in additional delays and opportunities for Saddam Hussein to withdraw peacefully from Kuwait.

Despite the growing concern at high diplomatic levels, it was not until January 20 that the possible environmental consequences of a war in Kuwait and Iraq jostled the complacency of many U.S. citizens. The start of the air war had dashed the hope that ecological concerns would prevail in George Bush's confrontation with Saddam Hussein. That evening, the television program *60 Minutes* aired a segment featuring Toukan and Sagan. In that broadcast, and again two days later on the late-evening news program *Nightline*, Sagan highlighted the possible effects that soot from the threatened oil fires could have throughout the Northern Hemisphere and raised the projected quantity of burning oil from 3 million to 10 million barrels a day. Many scientists subsequently scoffed at the higher figure, but Sagan was vindicated when the Kuwait Oil Company was able to assess the full extent of the disaster. He was essentially accurate on quantity but erred in assuming that it would all burn.

Sagan compared the soot that might rise above the troposphere through the process of self-lofting to the volcanic ash that exploded into the atmosphere from the eruption in 1815 of Tambora, a volcano on the Indonesian island of Sumbawa. That event was possibly the most powerful volcanic cataclysm on earth in the last 10,000 years. About one third of the mountain, which had been 13,000 feet tall, was obliterated in a blast more than a hundred times as powerful as the 1980 eruption of Mount St. Helens in Washington State. The superheated ash and gases released in the Tambora eruption rose quickly through the troposphere and brought on "the year without a summer" in Europe and North America. Harvests failed, and famine and riots ensued. It was an extreme example of the possibilities that could unfold if Kuwait's petroleum riches went up in an uncontrolled inferno, and it stayed in the minds of many people who viewed the broadcasts. The more probable outcomes—acid rain and a choking fallout of soot limited to the Arabian Peninsula and southern Asia—did not evoke the same emotional response.

Sagan's television appearances prompted the release of a number of other studies, funded by U.S., U.K., and German government agencies, which contradicted his scenario of an environmental catastrophe engulfing the entire Northern Hemisphere. Although some of these studies arrived at conclusions that generally confirmed Toukan's earlier prediction of a devastated environment throughout the Gulf region, and although the U.S. government was seriously concerned about the global impacts of massive oil-well fires in Kuwait, the overall sense was that the environmental threat had been exaggerated. Acting as if a war-engendered environmental catastrophe would be the public-relations equivalent of body bags, the Bush administration did not shrink from trumpeting studies that focused on disproving the gravest aspects of Sagan's warning.

Prominent among them was one conducted by Richard D. Small, of the Pacific-Sierra Research Corporation in West Los Angeles, for the Pentagon-based Defense Nuclear Agency. Small assumed that the flow of oil from exploded Kuwait wells would not exceed the rate at which oil had been produced in Kuwait proper before the Iraqi invasion: 1.59 million barrels a day. Like Toukan, he imagined that the entire flow would burn, but whereas Toukan thought that about 10 percent of the oil and gas escaping from the blown wells would be converted into soot, Small took 7.3 percent as his "smoke-emission factor," citing various controlled studies of crude-oil fires. His combination of a very small quantity of fuel, a modest smoke-emission factor, and an assumed maximum height of 10,000 feet for the smoke plume led him to conclude that the fires would have no effect on the earth's climate. Smoke from his theoretical fires wouldn't even shade the earth enough to cause noticeable cooling, according to his calculations. "In fact," Small wrote, "because the smoke remains at low altitudes, a slight warming may result."

Even though Small admitted that "the potential exists for an environmental catastrophe," the environmental organization Greenpeace attacked his work as wrong in nearly all its predictions and expressed dismay at the credibility it had attained, thanks to the support the White House had given it. Toukan maintains that the more spectacular aspects of his and Sagan's predictions were blown out of proportion by television producers, but Small contends that even the hint of a parallel with an event like the explosion of Tambora was irresponsible. The more environmentalists cry wolf, he said, the less credibility their voices will have.

Small published his study in the respected scientific journal *Nature*, which presented it as a "commentary." He failed to include the fact that practically all the 363 wells producing oil in Kuwait when Saddam Hussein's tanks arrived were doing so by means of subterranean pressure. Instead he digressed to discuss the gas-injection technology Kuwait used to protect its oil-bearing formations. He indicated that, without the injection of gas, the flow of oil to the surface would probably decrease and further suggested that even if the Iraqis exploded all the wells in Kuwait, the "worst-case" flow would not exceed 2 million barrels a day. He dismissed the possibility that the fires would merge, as they later did, into blazing rivers and lakes that allowed the smoke plume to rise higher than the plume from "point-source" fires. Small assumed that the fires would burn more efficiently than Toukan had predicted and would therefore emit less smoke and more carbon dioxide. Specifically, his worst-case calculations showed that on the average every barrel of oil burned would inject 10.2 kilograms of smoke and 109 kilograms of carbon dioxide into the air. Over the course of a year's burning at 2 million barrels a day, the totals would be 7.43 million metric tons of smoke and 79.3 million metric tons of carbon dioxide, 45 percent and 67 percent of what Toukan had calculated.

Another study, conducted by the Albuquerque-based Sandia National Laboratories for the U.S. Department of Energy, took the ignition of 720 oil wells as its worst case. Sandia researchers concluded that each blazing well would effectively be a blowtorch in which combustion would be very nearly complete, resulting in still lower amounts of soot and little or no fallout of oil droplets. The Sandia study turned out to be quite accurate in predicting combustion efficiency but far off the mark in its optimistic outlook for oil-droplet fallout.

Once the oil began flowing down the Saudi coastline, and the first, horrific images of the oil-well fires reached television audiences, few doubted that the Gulf War's ecological consequences would be felt the world over. One of the first scientific assessments of those consequences began, "The question of whether the smoke from the burning oil wells in Kuwait presents a serious threat to global climate has been the subject of considerable scientific debate and widespread public concern." Yet the Department of Energy explicitly forbade its scientists to discuss their research with the media. In an internal department memo obtained by *Scientific American* senior writer John Horgan, public-information officer John Belluardo specified the language that department scientists and contractors were authorized to use in public discussions of the "environmental impacts of the fires/oil spills in the Middle East." They were allowed to say: "Most independent studies and experts suggest that the catastrophic predictions in some recent news reports are exaggerated" and "We are currently reviewing the matter, but these predictions remain speculative and do not warrant any further comment at this time."

Nevertheless, the U.S. government took the global threat posed by the oil-well fires very seriously. Even after William Reilly, administrator of the Environmental Protection Agency (EPA), returning from a visit to postwar Kuwait and Saudia Arabia, said that he was "encouraged" by the environmental situation there, the Department of State called for airborne research to determine "the risks from the oil fire smoke to human health, the environment, crop productivity and the global climate," according to a department cable obtained by Horgan.

With the transportation and communication systems of Kuwait in ruins after the war and access to the Gulf region severely restricted, months passed before definitive studies on the composition and distribution of the smoke plume could begin. Until as late as May 1991, the best information available on the oil fires' environmental effects came from researchers operating outside the United States. In the absence of hard data on the plume's composition, they were limited to the use of remote-sensing technologies and computer models to estimate whether such calamities as disruption of the monsoons or further destruction of stratospheric ozone might follow in the wake of the fires.

A group from the British Meteorological Office, operating on the assumption that about 1.5 million barrels of oil were burning per day, issued a gloomy prognosis for areas near the fires but concluded that "the smoke plume is very unlikely to reach the stratosphere, and the bulk of the smoke will remain within the lower troposphere before being deposited on the ground within a week or so of its emission." The researchers determined that, within 120 or so miles of the fires, the smoke plume would be so dense as to reduce daylight "to near nighttime levels" and cause reductions in the daily maximum temperature of roughly 18 degrees Fahrenheit. On the subject of what effect the smoke might have on the summer monsoon in Asia, they did not see significant changes beyond the natural variations from year to year. They found that decreases in rain or snowfall over southern Asia were unlikely but noted that the possibility of "increases in precipitation in certain regions of the continent." This statement was the first hint that smoke from the oil-well fires might have been involved in the tropical cyclone that struck Bangladesh on April 30, 1991.

Early on, environmentalists began attacking the ill-coordinated attempts to deal with the disaster that was rapidly spreading across southern Asia, and they have continued to claim that the U.S. government deliberately tried to downplay the environmental damage of the Gulf War. Friends of the Earth proposed a tax on petroleum to pay for cleanup and restoration efforts in the Gulf region and stressed that the "scope of the fires is greater than generally reported and . . . getting worse." It contrasted its own observations of the smoke plume's maximum height at 16,000 feet with NOAA and EPA reports of between 8,000 and 11,500 feet. In the view of a Greenpeace writer, the U.S. government discouraged research into the war's ecological consequences because "the Bush administration is vulnerable on environmental issues," and deep probing of the war's environmental legacy "may bring more scrutiny at home."

Time has taken the environmentalists' criticism of the availability and accuracy of data on the smoke plume and its effects out of the limelight, even though many of their concerns appear to be well-founded. But the prospects for large and coordinated restoration efforts remain slim. While the fires burned, satellites constantly monitored their continent-spanning trail of smudge. Weather stations in Asia, across the Pacific, and in North America were on the alert for smoke particles that might have risen from the fires in Kuwait. And airborne meteorological laboratories penetrated the plume on a number of occasions and sampled its contents. By the end of 1991, researchers were beginning to discuss some of their findings at scientific meetings, and shortly after that, the findings appeared in the scientific literature, where standards of accuracy and comprehensiveness tend to be high. The U.N.-affiliated World Meteorological Organization in Geneva, and the National Center for Atmospheric Research (NCAR) in Boulder, Colorado, have both established data bases for information on the fires' effects, which is freely available to academic and environmentalist researchers alike.

Beginning in mid-May 1991, researchers from NCAR and the University of Washington in Seattle conducted about two hundred hours of sampling in the plume over the course of four weeks, the most thorough sampling at the time when the fires were raging at near-peak ferocity. Led by the University of Washington's Peter V. Hobbs and NCAR's Lawrence F. Radke, the scientists flew thirty-one missions in two meteorological-research aircraft, weaving back and forth through the plume at various altitudes in accordance with courses plotted by computer. One of the first things they noticed was that large numbers of fires billowed forth white smoke, which some observers had taken to be water. Indeed, the smoke rising from the fires came in every shade of gray; fires from natural-gas wells burned with almost no smoke at all, and burning oil lakes gave off the thickest, blackest smoke. Reporting on measurements taken by the University of Washington, Hobbs and Radke stated that the white-ishness in the lighter-colored plumes came not from water, which did rise with the oil from rocks below, but from the salt that had dissolved in the water. The water in the oil boiled away immediately in the heat of the fires, dissolving invisibly into the dry desert air over Kuwait and leaving behind the salt particles.

On the basis of their sampling, they determined the total

amount of carbon that the fires were releasing. Since they knew the carbon content of the unburned oil, they could calculate precisely how much oil was being burned at the time they collected their samples. The figure they arrived at was about 4.6 million barrels a day, which correlated with the Kuwait Oil Company's estimate of an average daily loss of nearly 11 million barrels and with other scientists' observations that only a third to a half of the gushing oil was burning.

They also found, significantly, that the fires were burning far more efficiently than anyone, except the researchers at Sandia Laboratories, had foreseen, producing far less soot and far more carbon dioxide per barrel than Toukan, Sagan, or Small had predicted. The University of Washington's measurements showed that the burning of an average barrel of oil injected 0.74 kilograms of soot and 400 kilograms of carbon dioxide into the air—5 percent and 370 percent of Toukan's predictions, respectively. If the fires had burned at a rate of 4.6 million barrels a day for a year, the University of Washington's sampling shows that the atmosphere would have been burdened with 1.24 million metric tons of sooty particles—7.5 percent of what Toukan had predicted—and 670 million metric tons of carbon dioxide, more than $5\frac{1}{2}$ times his prediction.

Alarming as the numbers appear, they conclusively show that the Northern Hemisphere was not imperiled by the threat of nuclear winter. The figure of 1.24 million metric tons is only about 1 percent of the total amount of soot involved in a nuclear winter, and only a small fraction of that 1.24 million metric tons would have been aloft at any single point in time. Even with the vast carbon-dioxide emissions that Hobbs and Radke found, their measurements—seen also in the context of the fast work of the fire fighters—put to rest, as well, the question of whether the oil fires would contribute significantly to greenhouse warming. Even if they had burned at that furious rate for a year, the carbon dioxide emitted would have been only about 3 percent of the world's total emissions from fossil-fuel burning in 1989.

Nevertheless, the findings of Hobbs and Radke are far from comforting. The rate of soot emissions they reported was "thirteen times greater than soot emissions from all combustion sources in the United States" and "equivalent to the soot emitted by about three million heavy-duty diesel trucks being driven at 30 miles per hour." The rate of sulfur-dioxide emissions from the fires was "57 percent of that from all the electric utilities in the United States."

Hobbs and Radke further found that the smoke plume was substantial enough to create its own weather systems. The heat of the smoke particles as they rose from the fires and as they were subsequently heated by sunshine boosted them well into the troposphere, usually to altitudes of between 10,000 and 16,000 feet, but at times to almost 20,000 feet. As the plume streamed out of Kuwaiti airspace, the air to either side of it was swept along in its flow, according to NCAR's measurements, which showed that winds converged toward the middle of the plume, increasing in speed as they did so. The smoke was so dense that, for hundreds of miles downwind, it prevented much of the sun's energy from reaching the desert surface. With the air near the ground failing to warm up as it usually did, it could not rise and disperse; rather, it remained trapped, with all its pollutants near the surface. The situation in Kuwait might have been worse, however, if more of the sun's energy had warmed the "surface layer" of air, causing it to rise. Such movement would have allowed the surface layer to mix with higher air and could have brought more of the plume's heavy load of pollutants in contact with the residents of Kuwait. As it was, people living under the pall of the smoke plume were usually protected from its worst threats. The price they paid was a black sky overhead and, depending on how close they were to the fires, increasing amounts of fallout in the form of soot and oil droplets.

Researchers flying through the plume often wore gas masks to protect them against not only the smoke but the unhealthy levels of sulfur dioxide. Interestingly, University of Washington scientists found that at least half the sulfur dioxide and nitrogen oxides emitted by the fires was converted into particles in about one hour. Other researchers have suggested that the chemicals that cause acid rain may have somehow combined with the plume's salt, which was responsible for about 30 percent of the mass of the sooty particles, and thereby been transformed into chemicals that Hobbs and Radke were not capable of measuring. In any event, the unexpected "loss" of the sulfur dioxide and nitrogen oxides points to the puzzling nature of the chemical reactions that took place in the plume. Thousands upon thousands of organic molecules, many of them toxic or otherwise hazardous, made up more than 2 percent of the plume. Little effort has gone into characterizing this mix beyond labeling it "total organic carbon in particles or vapor."

By concentrating on a few dangerous chemicals known in Northern industrial cities as "priority pollutants," researchers investigating the smoke from Kuwait's oil-well fires were able to allay some fears about the climatological and public-health implications of the oil fires, but they failed to take into account the complexity of the mixture of chemicals and particles in the plume. It was constantly changing and evolving as it spread outward from Kuwait. Several million dollars was spent to collect this sort of information during the NCAR–University of Washington airborne research, and scientists are ready to analyze those data and so unlock some of the plume's chemical mysteries. But government agencies have failed to fund the work. In Kuwait, public-health scientists deplored the limited amount of useful information released by foreign researchers. "Imagine all the toxic chemicals that could be present in tons of smoke from crude-oil fires," said Sami al-Yakoob of the Kuwait Institute of Scientific Research. "We will never know the poisons that we were exposed to."

The University of Washington did, however, collect information that will be used to correct the most recent round of nuclear-winter calculations, which appear to overestimate the amount of smoke that burning crude oil produces. The calculations assume that about 6 percent of the mass of burning crude oil is converted into soot, but Hobbs and Radke reported that, even in the blackest smoke from the ground fires and burning oil lakes, the figure was less than 3 percent. Another discovery was that sooty particles from Kuwait absorbed water avidly. Previously, scientists had thought that soot from oil fires would, like oil, not mix readily with water. Because the sooty particles from Kuwait did absorb water, meteorologists now recognize them as efficient cloud-condensation nuclei, or CCN, particles that serve as tiny platforms on which water vapor dissolved in the air can condense and form clouds.

Because the particles were efficient CCN, Hobbs and Radke thought that they would be quickly washed from the air by raindrops that condensed around them, and, indeed, reports of black rain came from all areas of the Middle East during the time the fires raged. Black snow was reported as well, far away in the Himalayas. But CCN cannot form clouds in the absence of water vapor, and the heaviest rainfall ever recorded in Kuwait was only 0.15 inches in twenty-four hours. The particles would, then, remain in the air and possibly travel great distances before encountering air moist enough to form clouds. Such air is characteristic of the Asian monsoons.

By the time smoke from the fires had traveled 1,000 miles, it had dispersed to the point where satellites could no longer detect it with cameras, but ground-based meteorological stations as far as halfway around the world from Kuwait subsequently detected soot that appeared to have been engendered by the Gulf War. The first such evidence came from the National Oceanic and Atmospheric Administration observatory on the Hawaiian volcano Mauna Loa. In early February, the observatory detected a "spike" of soot, a sudden rise and fall in the air's soot content. The timing suggested that large oil fires had to have been burning weeks before Saddam Hussein's troops began detonating Kuwaiti wells. The only other cause of such fires during the war was Coalition attacks on Iraqi oil-storage facilities.

From May to July 1991, Douglas Lowenthal, an atmospheric chemist, collected additional air samples both at the mountaintop observatory and at sea level on the island of Oahu. He analyzed them carefully and discovered that soot from oil-well fires in the Middle East was traveling the more than 9,000 miles to Hawaii. He concluded that his samples were evidence for the possible influence of the Kuwaiti oil fires at Mauna Loa and Oahu. The concentrations of soot in his samples were so low, however, that he believed the soot would have no effect on the world's climate.

NOAA scientists have found unusual amounts of soot in the air over Wyoming, but what is more significant is the altitude at which the soot was collected—more than 35,000 feet, which means that it lies in the boundary between the troposphere and the stratosphere. Again, however, scientists insist that their detection of soot at that altitude says more about the sensitivity of their instruments than about an additional threat to the ozone layer, which lies in the stratosphere. The concentration of soot was low enough, and the fires were put out quickly enough, they say, to ensure the safety of the Northern Hemisphere's climate from the influence of the Kuwait oil fires.

Before the fact, all researchers agreed that Kuwait's environment would be devastated and that the ecological damage would diminish with increasing distance from the fires. In his commentary in *Nature,* Richard Small predicted correctly that the pollution would fall out on a wide swath across southern Iran, Pakistan, and possibly northern India. "In the worst case," he wrote, "human populations would be exposed to persistent air pollution; grazing herds may be affected by accumulated ingestion of soot; agricultural and ecological systems impacted by a covering of soot; and water systems polluted by the fallout and accumulations in run-off." Ecological damage caused by the oil fires is increasingly difficult to prove the farther from Kuwait it occurred. And even in Kuwait, researchers have to take the correct measurements in order to prove whether some of the damage they saw with their own eyes was caused by the fires.

CHAPTER SIX

The Smoke Seen Round the World

≶ ≶ ≶ ≶ ≶ ≶ ≶ ON APRIL 30, 1991, a devastating cyclone hit the impoverished nation of Bangladesh, which occupies the low-lying plain at the delta of the Ganges and Brahmaputra rivers. Raging out of the Bay of Bengal with 150-mile-an-hour winds, the storm drove ashore a surge of water towering twenty feet higher than normal high tides. Lands many miles inland were flooded, and between 140,000 and 200,000 people died.

During the latter part of June, pelting rainstorms hit areas of central and eastern China, causing catastrophic floods in the valleys of the Yangtze and Huai rivers. Press reports out of China stated that the floods affected 200,000,000 people—a fifth of the country's population. Anhui Province, east of Shanghai, was particularly hard hit. In all, about 3,000 people died in the floods.

The winter of 1991–92 brought bizarre weather to Israel, Jordan, Lebanon, and Syria. Blizzards, unheard of in the lifetimes of most of the inhabitants, pummeled the eastern Mediterranean region repeatedly. Swollen rivers unearthed hundreds of land mines that Israelis and Arabs had planted, endangering the lives of everyone living downstream. So much rain and snow fell that Israeli officials had to open floodgates on the Sea of Galilee to prevent the waters from spilling over its shores and flooding nearby cities and towns, thus wasting a precious resource in the chronically drought-stricken area. For the first time in history, Israel's chief rabbi encouraged the faithful to pray for an end to rain and snow. According to press reports, hundreds of people in the region died in floods, avalanches, and the collapse of snow- and ice-covered buildings.

In a public broadcast on March 29, 1992, Kathy Sullivan, a mission specialist aboard the space shuttle *Atlantis,* reported with amazement that the earth's lower atmosphere appeared to be choked with a mantle of dust and smoke. "I've really been struck on this flight by how hazy the lower atmosphere appears to be," she said from almost 200 miles above earth. "I flew just about exactly two years ago, [and] I can't believe [the deterioration] is seasonal, because this time of year two years ago it was lots different from this."

In her reports from the *Atlantis,* she never mentioned the oil fires in Kuwait, which had been extinguished four and a half months earlier, but she did note that she could see smoke from regions where farmers were using fire to clear their land. She and her companions aboard the shuttle also saw especially large amounts of dust in the air over northern Africa. According to NASA, which said that the primary task of the mission was atmospheric research, particularly into how natural and human influences affect the planet's gossamer envelope, instruments on the plane also found evidence in the stratosphere of ash from the June 1991 eruptions of the Mt. Pinatubo volcano in the Philippines.

Although rainfall is sparse in the eastern Mediterranean region, and snowfall even sparser, the freak winter of 1991–92 was not unprecedented. Meteorologists cited in press accounts of the unusually stormy winter said that it was normal for the region to experience such a winter once every fifty to one hundred years. None of the reports, however, addressed the issue of whether the explosive events that took place in Kuwait and Iraq might be linked to the strange weather, even though scientific investigators had established the possibility during the second half of 1991.

To be sure, typhoons in Bangladesh and floods in China are tragic facts of life in those countries. In November 1970, 300,000 people in Bangladesh perished in a storm; in May 1985, a typhoon took the lives of 10,000 people there. In July 1981, more than 1,300 Chinese were lost in a flood that swept Sichuan. In northern China, 200,000 were reportedly killed in floods in 1939; an incredible 3,700,000 perished eight years earlier when torrents raged down the Huang He River. China and Bangladesh are poor, overpopulated countries, where millions live with virtually no protection from unusual or unexpectedly violent weather. When a natural calamity strikes, a much higher percentage of the population is at risk than in developed countries, where flood-control and emergency-shelter systems are well developed.

No scientist can claim with certainty that smoke from Kuwait's oil wells caused the cyclone in Bangladesh, the floods in China—or the snows in Jerusalem, Damascus, and Beirut, where dust kicked up by trench digging and military maneuvers, rather than smoke, may have played a role in the region's strange weather. But neither can anyone claim that it was impossible for the Gulf War to have been the agent that altered the weather at great distances from Kuwait. Establishing or disproving cause and effect is monumentally difficult. It is nevertheless a fact that the smoke had characteristics that made it a possible weather modifier, and government researchers who have tried to follow the fate of the smoke as it dispersed throughout the troposphere believe that it was present over the Bay of Bengal and China when the destructive storms raged in 1991.

When atmospheric scientists Peter Hobbs and Lawrence Radke found that smoke particles from the Kuwait oil fires readily absorbed water and therefore could have served as "seeds" on which water droplets and, eventually, storm clouds could form, they and other researchers expected that, in the absence of moisture in the air over the Gulf region, the prevailing flow of air out of the region would carry the lighter particles away. Throughout 1991, that flow took the smoke in the lower atmosphere south over the Gulf, then eastward across the Arabian Sea, southern India, the Bay of Bengal, southeast Asia, and finally out across the Pacific, dispersing and becoming harder to detect all along the way. "We lost our ability to detect the particles at about 200 miles downwind," Hobbs reported. He believes that both dispersion and fallout are the reasons the concentration of smoke particles in the air dropped to below his ability to detect them. "The heavy particles fell out pretty quickly," he explained.

As the wispy particles of smoke drifted eastward, they naturally became a smaller and smaller fraction of the total amount of any particulate pollution in the air wherever they were. The longer this dispersion and dilution continued, the more difficult it became for scientists sampling and analyzing the air at various locations around the world to identify Kuwait oil smoke in their samples. Nevertheless, scientist Douglas Lowenthal, with the samples collected at the NOAA observatory on Mauna Loa, and a handful of other researchers were able to do so. Lowenthal and his collaborators believed that the air they sampled arrived in Hawaii from a distant source because, during the weeks when samples were collected, the concentration of elemental carbon, or soot, rose and fell in parallel with that of other elements resulting from pollution. Furthermore, the samples Lowenthal took at sea level on Oahu exhibited the same unusual pattern. He therefore believed that the soot came from the oil fires rather than from the multitude of pollution sources in Asia, because the Asian pollution that is usually detected in Hawaii has a characteristic chemical "signature"—a particular ratio of the element vanadium to the element zinc—which was not present in the air sampled from May to June 1991.

Analyses like Lowenthal's are time-consuming and costly, and they require sampling stations far removed from pollution sources. Dispersed as the smoke from the fires was by the time it reached Hawaii, it was not dangerous in the way radioactive fallout is. Therefore, it was not reasonable to expect all of the world's most sophisticated air-sampling stations to go on high alert to track it. But it was a very interesting substance for atmospheric scientists; so they followed its path in the best way they could, by means of computer models.

Thomas J. Sullivan is a computer modeler at the Lawrence Livermore National Laboratories, operated by the Department of Energy in Livermore, California. The U.S. government uses his model of the atmosphere to predict where radioactivity from nuclear accidents is headed. His model tracked the paths of hazardous particles and gases released from Three Mile Island, Pennsylvania, in March 1979; from Chernobyl, in Ukraine, in April 1986; and from a number of other nuclear sites. In 1991, the government asked him to determine the likely path of the smoke plume and its process of dispersion as it moved from Kuwait across southern Asia and the Pacific.

According to Sullivan, the tempests that hit Bangladesh and China in 1991 were strong enough to be considered the types of storms that might strike once in a hundred years. "Both events were quite intense, statistically very anomalous," he said. His model showed that the weather systems of both countries drew in smoke from the oil fires. The identification of the soot particles as strong cloud-condensation nuclei was, Sullivan found, "very straightforward." No one doubted that under certain conditions a sufficiently dense concentration of soot particles could create a cloud out of moist air. Thus it was possible, Sullivan contends, that the smoke particles from Kuwait "contributed to the intensity of the rain" in Bangladesh and China.

In addition, the movement of the smoke particles over southern India, which Sullivan's model predicted, may have led to significantly abnormal seasonal weather patterns. In 1991, southern India had an unusually wet year, while northern India was unusually dry. The moisture that rain clouds drop on India during the spring and summer months comes from evaporation in the Indian Ocean. When central Asia heats up each spring and summer, warm surface air rises, and the moisture-laden air over the Indian Ocean sweeps up from the south, across India and over the Himalayas, to replace it. If Sullivan's model determined correctly that a stream of smoke particles constantly traversed the lower atmosphere of the southern half of the subcontinent, the moist air moving up from the Indian Ocean would have mixed with it—and possibly played a role in the distribution of excess rain in the south and the consequent lack of rain in the north.

Other atmospheric modelers around the world also determined that much of the smoke from the Kuwait oil fires would drift to the east, perhaps affecting the pattern of monsoon rainfall in significant ways. Representatives from the British Meteorological Office, the Max Planck Institute for Meteorology in Hamburg, Canada's Atmospheric Environment Service, Météorologie Nationale of France, and the Royal Netherlands Meteorological Institute met in Geneva in April 1991 to discuss their findings. They agreed that the oil fires gave off an enormous amount of soot—two-thirds of the total worldwide soot emissions during the months when the fires were raging most furiously, according to Joyce E. Penner of the Lawrence Livermore National Laboratories. They also agreed that the soot was certain to have a profound effect on the weather in the Gulf region and exert further influence over weather patterns in ways that would be increasingly difficult to predict or trace as the distance from the fires increased. The Canadian group stated that the smoke would result in "increases in monsoon precipitation over India."

The report issued by the Geneva meeting declared that the deposition of soot in the Himalayas would probably lead to faster melting of winter snows, and it supported the thesis that the smoke "might somewhat intensify the monsoon rain." It specifically cautioned, however, that "natural variability [in the monsoon rains] is so large that the detection of a significant change would be difficult." Rumen D. Bojkov, chief of the environmental division for the Geneva-based World Meteorological Organization and convener of the conference, later underlined this caveat by saying that "there is absolutely no evidence" of Kuwait oil-fire smoke causing the cyclone in Bangladesh.

Peter Hobbs, who is more confident with actual measurements than with the output of computer models, is skeptical of both the claims that the smoke had no influence in the disasters in China and Bangladesh and the claims that it triggered the terrible storms. "The atmosphere is enormously big," he says, "and it is quite difficult to change weather systems that have tremendous energies associated with them." He points out that, despite decades of research on weather modification, there is "no proof that one can change the precipitation that a storm produces on the ground by seeding clouds with artificial materials." The processes that go into the creation of a rainstorm are so complex, and each storm is so different from every other, that meteorologists cannot pinpoint the impact one subtle component may have.

"It's like anything else in science," Hobbs says. "Show me the evidence, and if it's convincing, I'll probably believe you." He points out that the normal amounts of air pollution in eastern and central China would have swamped any effects of smoke from Kuwait, and asks, "Why should that one particle from Kuwait have such a big influence if the many others did not?"

But according to Thomas Sullivan, the real problem is that meteorology has not yet advanced to the point where one can construct a convincing argument that Kuwait oil-fire smoke caused either of the deadly storms. "We still have a lot of unknowns about the triggering of precipitation by these soot particles," Sullivan says, "and I'm concerned that not too many people seem to be studying that now." A possible reason, in his opinion, is that "there has been a shortage of research funds, and since the smoke did not directly affect the United States, it was not a high priority for funding."

Even so, the World Meteorological Organization found enough money in 1991 to establish four new stations to monitor air quality, with a focus on pollution from oil fires. But it will be years before the data collected at the stations will be sufficiently scrutinized to determine what effects, if any, the oil fires had at great distances from Kuwait. As the years pass, the stations will continue to collect samples in order to ascertain a baseline of air pollution around the world. At the same time, the haze that shocked Kathy Sullivan from aboard the *Atlantis* could well continue to build and confound the process.

CHAPTER SEVEN

The Toll on Plants and Wildlife

ℰ ℰ ℰ ℰ ℰ ℰ ℰ N̲o ONE WHO WAS in the Gulf region during the summer of 1991 had any doubts about air quality. From Baghdad in the north to Yemen in the south, from Riyadh to far beyond the eastern slopes of Iran's Zagros Mountains, one saw a sky smudged by smoke and smelled the odor of burning Kuwaiti crude oil. The smoke was so dense and so continuous that locations more than 300 miles from Kuwait experienced their coolest summer on record.

Surprisingly, the sooty air did not appear to acutely threaten the health of people and animals living outside Kuwait. Niza Khan, an environmental specialist for Saudi Arabia's Royal Commission for Jubail and Yanbu, said that he was "amazed and could hardly believe" the low levels of pollutants other than soot in the air over al-Jubail. His department maintained an air-quality monitoring program throughout the time the gloomy pall persisted, assessing every hour the levels of such "priority pollutants" as sulfur dioxide, nitrogen oxides, ozone, particulates (including soot), and a grab bag of dangerous chemicals known as nonmethane organic carbon. In the presence of sunlight, these chemicals react with the oxygen in the air to produce ozone—which, despite its role as a blotter for the sun's cancer-causing ultraviolet radiation 40,000 to 80,000 feet, or seven to fifteen miles, above the earth's surface, is a toxic chemical for plants and animals when present in large concentrations near the surface.

During the months in which the fires exhaled the greatest quantities of pollutants, Khan found that in al-Jubail, only particulates and nonmethane organic carbon ever exceeded the air-pollution standards of Saudi Arabia, which he described as "stringent." The pollutants exceeded the standards on only fifteen days.

The commission never issued any alerts about the high levels of nonmethane organic carbon, Khan explains, precisely because its chemicals are known to be dangerous only in the presence of strong sunlight, and sunlight was invariably dim on the days when nonmethane organic carbon levels were high, because those were also the days when the load of sooty particulates in the air was at its highest. The production of ozone, the real danger, never rose to a threatening rate. "God protected us here in al-Jubail," Khan says, rolling his eyes heavenward, "but the people living in Kuwait unquestionably endured unhealthy amounts of pollution in their air."

Passing over the Gulf, the smoke plume from Kuwait dropped tons of soot particles and oil droplets into the water. At times, this fallout coalesced into a dark, mottled sheen covering several square miles until it slowly disappeared, mixing with the Gulf waters or evaporating. The sheens of well-fire fallout ended when the wells were all capped, but other sheens, moving out from heavily contaminated sediments along Saudi Arabia's coast, continued to spread onto Gulf waters for more than a year after the first oil was spilled. The Royal Commission had to keep an eye out for these residual oil sheens and have its response crews ready for the times when the sheens drifted dangerously close to cooling-water or desalination plant intakes. When they did, crews corralled the sheens with boom and collected the oil with skimmers.

According to MEPA, efforts to protect important environmental and wildlife areas along the coastline by and large failed, because "most of the available response equipment in Saudi Arabia had been allocated to the protection of industrial facilities." Greenpeace put it more bluntly: "No effort was made to protect environmentally sensitive areas." In a report on the Gulf War's ecological aftermath, Greenpeace went on to say that "this policy of sacrificing the environment for industrial protection contributed to the massive damage to the coast." As of the summer of 1992, only small attempts have been made to remove the oil (between one and two million barrels—four to eight times the *Exxon Valdez* spill) from shoreline sediments or to clean up the shoreline in any way. MEPA, which is openly and, according to Greenpeace, justifiably, skeptical of some of the methods used in the three-year *Exxon Valdez* cleanup, has, however, instituted an experimental program to develop alternative techniques to assist or hasten the natural cleaning processes that have now begun.

To assess the situation along the shores of its Eastern Province, the Saudis sent a twelve-man team into the field during the latter part of May 1991. Traveling along the more than 450 miles of coastline between the Kuwait border and Abu Ali, except for areas of impassable terrain, the team— whose members came from MEPA, King Fahd University of Petroleum and Minerals in Dhahran, Saudi Arabia's National Commission for Wildlife Conservation and Development, the European Commission, the U.S. Coast Guard, the Saudi Arabian subsidiary of the Bechtel Company, and Crowley Maritime Corporation of Seattle, Washington—completed the most comprehensive assessment of the spill's damage to date. The group divided the coastline into 147 sites, noting their physical characteristics, degree of oiling, and economic or archaeological significance, ranked them in order of cleanup priority, and recommended specific cleanup techniques.

Three months later, Greenpeace surveyed the spill's devastation along the Saudi coast as part of a regional investigation of the war's environmental aftermath. The Greenpeace assessment, which included areas in Bahrain, Kuwait, and Iran, covered fewer sites in Saudia Arabia than the Saudi shoreline assessment did, but examined those it chose in greater detail than the team assembled by the Saudis had. Representatives of the National Commission for Wildlife Conservation and Development and MEPA also participated in the survey, which Greenpeace was able to carry out partly because of its status as a nongovernmental organization that sends delegates to meetings of the International Maritime Organization.

Greenpeace determined that natural cycles of shoreline erosion and deposition during the months following the Gulf War gave rise to the nearly continuous presence of large areas of sheen in the northwestern Gulf. Tidal currents and seasonal storms constantly rework the arrangement of coastal sediments, picking up loads of sand and strewing them downstream. Greenpeace found that most new slicks slid out to sea when near-shore currents stirred up an area of heavily oiled sand or mud, usually a spit or headland between bays. Surprisingly, these heavily oiled areas were often not obvious but hidden under clean sand that currents had dropped on top of the contaminated sediments. In fact, some of the sites that Greenpeace selected for sampling, on the assumption that they were relatively untouched by the massive pollution, turned out to be oiled beaches that had been covered by recently deposited sand. In some places, the layer of clean sand was less than half an inch thick, and a hand spade was all that was needed to reveal the oily sediments below. In other areas, the oiled sediments were buried under fifteen inches or more of clean sand. In either case, the researchers often found that the layer of oiled sediments was quite thick— sometimes thicker than the team was prepared to measure. According to Greenpeace, the oil often bound sediments together in a kind of asphalt that resisted the power of currents to redistribute the sand and soil particles, thus inhibiting the oil's dispersion and natural breakdown.

Paul Horsman, one of the leaders of the Greenpeace expedition, found that the oil's impact on the shore depended on the physical characteristics of the site. "The north- and east-facing inlets suffered the worst pollution," he explained. "Particularly in Manifa and Musallamiya bays, oil covered the entire intertidal zone and extended from above the high-water mark to below the low-water mark and out as far as areas where the water was one to two meters [three to six feet] deep."

Six months after the oil began drifting south from Kuwait, it still remained fluid as it clung to subtidal sediments. When Greenpeace researcher Dale Roston sifted core samples she had taken from subtidal bottoms, shaking a sieve into water about three feet deep and letting the sediment drift off in the current while she looked for shells and worms in the sieve, the oil-coated particles of clay billowed away, the ironic mirror image of the smoke clouds billowing up from Kuwait's oil fields.

As might be expected, the oil percolated deeper in coarser sediments. In the vicinity of Ras al-Mishab, a peninsula about halfway between al-Khafji and Saffaniya, Horsman and his companions discovered sheltered areas where the oil had sunk into coarse sediments of sand and broken shell, which allowed it to penetrate to depths the researchers were not capable of measuring. Wading or maneuvering inflatable dinghies through near-shore waters, they found large patches of dead sea grass near the sheltered areas of coarse sand and broken shell. Rocky shores, which along the Gulf coast are generally made up of large, flat rocks, wore a coat of tar that made them look like broken-up parking lots. In areas of mud and fine sediments, the oil formed an impenetrable layer covering large expanses of the algal mat that is basic to the Gulf's biological productivity. The layer of algae and bacteria probably prevented much of the oil from penetrating as deeply as it did in less fine sediments, although, according to the report, "animal burrows acted as conduits for the oil to percolate into the sediments."

Many stretches of the Saudi coast fell victim to the oil spills during a period of especially high tides and onshore winds at the beginning of spring in 1991. These carried the oil to areas farther up the shore than normal tides reach. Oil from Kuwait also rode those high tides along tidal creeks and left its smear as far as three miles inland, reaching habitats where plants and animals can survive only by being extremely tolerant of salt in their tissues. Horsman and his group came to the conclusion that the Saudi coast suffered dramatically from the oil spills.

The predicament of the Saudi coastline was unique in the history of even the worst oil spills, according to Greenpeace. A combination of the inability to control the source and the lack of effective protection for shorelines north of al-Jubail led to the 400-mile-long catastrophe. In "normal" oil spills, the volume of oil is much smaller and cleanup efforts usually get under way rather quickly, preventing the contamination of continuous stretches of shoreline. In such cases, where there are areas of relatively clean shoreline interspersed among the polluted areas, recovery is aided by the colonization of the more sorely afflicted shoreline by species from the less damaged stretches.

In some of the hardest-hit stretches of the Saudi coast, Greenpeace found sediments containing up to 7 percent oil by weight, and "it was common to find sediments that were 1 to 2 percent oil." The oiliest beach was near a tidal inlet known as Brice lagoon, a north-facing sheltered flat lying along a stretch of shoreline about halfway between Manifa Bay and Ras al-Zawr. Greenpeace sampled the upper shore near Brice lagoon and found sediments containing from 6.5 to 9 percent oil. MEPA's Abdul-Jaleel al-Ashi, who participated in the earlier assessment, acknowledged that "the situation is very sad at Brice lagoon; we weren't able to protect it at all, and a really incredible amount of oil covered the whole area." Al-Ashi's colleague, Aziz al-Amri, who flew aerial reconnaissance over the coastline throughout the spill crisis, explained that "it's not surprising Brice lagoon is one of the oiliest areas. When the oil hit there, it was really thick and completely covered the water out into the Gulf for as far as I could see."

The assessment made by the Saudis in May 1991 also noted hitherto unknown shoreline contamination. The assessment team classified half of the 147 identified sites as bearing a "heavy" concentration of oil, which they defined as oil covering at least 50 percent of a shoreline in a swath at least five meters (sixteen feet) wide. If oil penetrated more than ten centimeters (four inches), they also classified as heavily contaminated those sites with less or narrower coverage. In its report, MEPA noted that the low levels of wave and tidal energy in the Gulf would result in a slow recovery for the coast: "Sensitive habitats such as salt marshes may not recover for several years," and "mangroves may take two or three decades."

Although hundreds of thousands of barrels of oil probably bypassed the Saudi coast altogether—soiling instead the shorelines of Qatar, Bahrain, and the United Arab Emirates in the form of sheen and tar balls—areas beyond the principal focus of the damage in Saudi Arabia "escaped much of the major impact of the Gulf War spills," according to the Greenpeace report. This does not mean, however, that the other shorelines of the Gulf are not victims of oil pollution. The Gulf is of course the world's busiest highway for the traffic of oil and petrochemicals, and many researchers estimate that in a year free of major oil spills, at least one *Exxon Valdez*-worth of oil enters Gulf waters in the course of normal drilling, production, and tanker activities. In some areas, the "normal" burden of oil pollution can be heavy. While surveying the coast of Bahrain, the Greenpeace team took a sample in the vicinity of an oil-refinery discharge at Manama. "The analysis revealed an oil content of more than 7 percent, which is clearly due to local pollution," Greenpeace reported.

The situation is no different on the other side of the Gulf, where Greenpeace sampled beach sediments during the latter part of September 1991 and found a number of sites along the Iranian coast where the sediments contained up to 4 percent oil by weight. The group attributed much of this, however, to other accidents and operational spills. "While it is possible that some oil from the recent Gulf War may have reached the [Iranian] coast," it said, "further work is required to identify the origins of this oil."

In Saudi Arabia, most of the contamination was in the intertidal zone, which includes ecologically important tidal flats. Although these salty and oxygen-poor habitats are inhabited by creatures that survive only because of their special adaptations to harsh conditions, they are among the most biologically productive ecosystems of the Saudi coast. Saudi Arabia's tidal flats are composed of very fine-grained sediments, which means that these areas are also the ones where the problem of removing large quantities of oil is most intractable. It is almost impossible to walk across parts of these tidal flats: with each step one sinks up to one's calf in mud that has the texture of potters' clay. There is no way that earth-moving equipment can operate in such a mire to remove the contaminated mud, and repeated flushings—either by the tide or by cleanup workers with hoses—rinse out only the topmost fraction of an inch. MEPA and the National Commission for Wildlife Conservation ruled out the use of heavy equipment for other reasons, however. First of all, removing contaminated sediments raises the disposal problem: where to put the oily mud? Second, as Abdulaziz H. Abuzinada, secretary-general of the National Commission, pointed out, "The use of heavy equipment would only compound the damage already caused to the fragile soil structure of these habitats."

Mangroves coated with oil received a large share of the media's attention to the plight of Saudi Arabia's Gulf coast, perhaps because the image of a mangrove stand symbolizes to a Westerner all the lushness and impenetrability of a tropical rain-forest at the water's edge of an island paradise. The image, however, has little to do with the handful of dispersed thickets of stunted black mangroves growing between the Kuwait border and Abu Ali. An expanse of mud and algal mat usually separates them from the low-water line, they rarely grow to heights of more than about five feet, and they present no barrier to travel farther inland since it's easy enough to slog through or walk around them. They represent only a fraction of the total of Saudi Arabia's Gulf mangroves, most of which grow along the shores of Tarut Bay, just south of Ras Tannura, and escaped the spill. But the northern stands play an important role in the ecology of the Gulf. Normally they stabilize sediments and gradually extend the coastal flat seaward. MEPA placed a high value both on the 5½ miles of mangrove stands and on the salt marshes usually found just upland, and gave them the highest priority for cleaning. According to Greenpeace's Horsman, however, the high value that MEPA claims to put on the mangrove stands has never been supported in Saudi Arabia, where industrial development has destroyed most of the mangrove areas present before the discovery of oil.

Gurma Island is ringed by the largest mangrove thicket in the area affected by the spill. It was also in the middle of Khaleej Marduma, where the million-barrel cleanup took place, the largest oil-spill cleanup operation in history. During the weeks when Jim O'Brien and his crews were working with their mud cats and "wemust" pumps, the plants growing closest to the low-water line were completely submerged in thick oil twice a day. Those growing a few feet higher up were oiled only at the base of their trunks, but even they suffered in the spill because oil had plastered their pneumatophores, or breathing structures. Pneumatophores serve as a kind of snorkel for the mangrove, rising through the soil from the plants' roots into the air, where they absorb oxygen for delivery to the roots. Mangroves evolved these structures because the extremely fine-grained soil in which the plants grow does not hold enough oxygen to sustain them. The twice-daily inundations of weathered crude oil during March 1991 effectively tarred over the mangroves' pneumatophores, sealing the trees off from the air. As O'Brien's and the other crews' efforts to collect floating oil wound down in April, an attempt to save the mangrove thicket by showering it with low-pressure sprays of seawater began.

The sprays rinsed off a heavy load, which workers collected with booms and skimmers deployed in near-shore waters. In April, when the plants began taking their seawater showers, they were literally dripping with weathered crude oil, and when I visited the island seven months later it was still easy to imagine what a mess it had been. Mangrove plants growing close to the low-water line were dying, their leaves yellowed and blackened and their trunks and branches coated with tarry oil. Even many of the plants growing in the higher and dryer areas still had pneumatophores clogged by oil; cutting off a pneumatophore and stripping it of its tarry bark often revealed dead tissue underneath, a bleak sign for the plant.

But the decline was not universal. Many of the plants sported new leaves, and some even cast off young plants in a process peculiar to mangroves: the seeds cling to their "mothers" until they are fully germinated, complete little plants with their first leaves and vigorous tap root. This phenomenon, like the pneumatophore, evolved to cope with the oxygen-poor soil in which mangroves thrive. Without young leaves to suck oxygen out of the air, the baby plants would be unable to establish themselves in the mud. When they finally drop from the mother plant, they float away with the tide and take root where they may. The absence of pools of standing oil on Gurma Island when I was there in November 1991 is encouraging for the baby plants that were germinating and that might one day replace the ones that did not survive the spill. But despite the work to rid the island of oil, strips of oil sheen still floated through the tidal channels, swirling around the baby mangroves. With luck, the young plants will prove tougher than the oily mud in which they were taking root. All the new growth could indicate the success of the spraying effort, but it is also possible that it is merely a physiological reaction to the extreme stress that the plants suffered from February to April, a reaction in which the plants' last reserves of nutrients and energy are burned up in an attempt to overcome the choking effects of the oil spill. Only time will tell.

MEPA gave high priority to the protection and cleaning of mangroves not only because the plants are rare and the most obvious component of the highly productive intertidal zone but also because they play a role in creating and maintaining an area of great biological activity and diversity. The small channels that meander through the mangrove thickets are nurseries for shrimp and fish, and innumerable crabs, clams, and worms make their homes in burrows dug into the compact soil that collects around the mangroves. There they collect bits of food brought in by the nutrient-rich tides. In turn, they attract huge flocks of migratory birds that swoop in for sustenance on their flights between tropical Africa and northern Europe or Asia. Although the oil that has sunk so deep into Saudi Arabia's sandy shores is a blot on what would otherwise be a marvelously wild and rarely visited seascape, and although it has proved fatal to the ghost crabs and beach fleas that once lived in these sandy areas, it is in the normally teeming areas of algal mat and mangrove that the oil's ecological repercussions will be felt the most.

According to Greenpeace, most of the muddy intertidal areas that line the sheltered bays of Saudia Arabia's northern Gulf coast had "no sign of surviving marine life" when the group surveyed the coastline in August 1991. The first evidence of life Paul Horsman and his companions saw during their expedition appeared far below the low-water mark. The few creatures they found living in the near-shore waters were usually representatives of a few species known to tolerate heavy pollution loads, animals like certain types of mussels and polychaete worms, which cement houses made of sand grains onto their bodies. Greenpeace did not suggest that its observations, made at only a limited number of sites, might be indicative of all the muddy intertidal areas attacked by the spill, but the severe damage they described is of the sort that could cripple the environment for many years, or forever. Each week since then, fortunately, has brought small, sometimes barely noticeable improvements, and just three months later there was at least hope that the most productive habitats would not become eternal deserts. Still, signs of improvement can be more apparent than real. Larval animals

that survived the crisis phase of the oil spill could subsequently suffer developmental abnormalities, shortened life
spans, and reproductive failure.

Steve LeGore, an environmental consultant to the International Maritime Organization and the director of Battelle
Ocean Sciences in Duxbury, Massachusetts, visited a number of sites along the Gulf coast of Saudi Arabia in November 1991. In his close inspection of the mangrove
pneumatophores on Gurma Island, he found that the ones
that had had the thinnest coating of oil often had living, green
tissue under their bark. "This shows that they have been
able to absorb enough oxygen through the oiled pneumatophore to keep the plant alive," LeGore said. He also saw
that several species of crabs and snails had begun to reestab-
lish themselves in heavily oiled areas that had been cleaned
with seawater rinses. "The recolonization by crustaceans is
significant," LeGore noted, "because they are especially
sensitive to hydrocarbons." He found further encouragement in the evidence that small fish known as gobies were
returning to their typical habitats on sandy bottoms that had
become relatively clean. The return of the goby fish is significant for several reasons, according to LeGore. First, the
fish are very territorial, which means that the environment
must meet some standard of cleanliness before they move in;
second, their return to Saudi Arabia's near-shore waters indicates a cross-phylum balance in the kinds of animals coming back to the oiled areas.

Greenpeace's discovery of large patches of dead sea grass
in subtidal areas where the oil had sunk deep into sediments
of coarse sand and broken shell is troubling because these
plants also contribute in important ways to the biological
productivity of the Gulf. Three species of sea grass grow in
Saudi waters, from just below the low-water line to depths
of more than thirty feet. The plants cover over four hundred
square miles, occupying a key position in the life cycle of

many species of commercial fish and shrimp. The sea grasses are particularly important because they, along with the coral reefs found in the Gulf, are the principal agents of converting the sun's energy into living tissue in bottom-dwelling communities. Normally, these communities depend on tiny plants called phytoplankton for their primary energy source. But in the Gulf, sea grasses and coral are important exceptions. They capture much more energy than they need to sustain themselves and so become a dependable food source, not only for the creatures that spend their entire lives in the sea-grass beds but also for those that carry the sea grasses' bounty to neighboring environments as well. Abdulaziz Abuzinada, of the National Commission, issued a guarded assessment: "preliminary studies indicate that the sea-grass beds and coral reefs largely remain intact, although it is premature to draw conclusions at this stage." Greenpeace supported the cautionary tone, saying that "larval development, reproductive capacity, and possibly the immune system" of surviving corals could have been harmed by chemicals present in the spilled oil.

The biological productivity of the intertidal mud flats and subtidal sea-grass beds culminates in the fish and shrimp that appear in open-air fish markets in the cities and towns dispersed along the Gulf shores. The most popular fish in the markets are barracuda, king mackerel, and two species known locally as *subaitee* and *hamour;* both Kuwaitis and Saudis are particularly fond of *hamour.* These are all large fish that feed on the Gulf's abundant bait fish—the silversides, sardines, and anchovies. There is also a rarer group, which includes impressive carnivores—at least ten species of shark and two varieties of bottlenose dolphin. Like the shrimp and the mole lobster that inhabit the Gulf's bottom, the bait fish depend almost entirely on the Gulf's primary producers. But all these animals could suffer if Saudia Arabia's sea-grass beds and intertidal flats fall into serious decline.

One indication of decline was the six-week delay in the opening of the shrimp fishing season off the Saudi coast. Finding that its early catches were underweight, the Saudi Fishing Company decided to leave the shrimp alone for the season. Subsistence fishermen also abandoned their trap fisheries along the Gulf shores of Kuwait and Saudi Arabia. Oil had ruined their nets and traps and damaged their boats, and they were afraid of mines and other unexploded munitions. Greenpeace reported evidence of fish die-offs in certain areas and cited the fisheries ministry of Bahrain regarding the discovery of fish with oil-induced blemishes. According to MEPA, shortly after the spill Saudia Arabia's fishery in the northern Gulf dwindled to zero, but by year's end fishermen reported that they were catching 20 to 25 percent of their previous numbers of fish, albeit in smaller sizes. This is a serious development if the smaller fish represent most of the spill's survivors, since the overfishing of an immature community can lead to long-term and possibly irreversible damage to the fishery. There is the possibility, however, that the larger fish are still present in the Gulf, but in deeper waters, which offer more plentiful food.

More definitive information on the Gulf's ecological health will come from a 100-day oceanographic cruise that began in February 1992. Aboard the *Mount Mitchell*, a research vessel operated by the U.S. National Oceanic and Atmospheric Administration, scientists from every country bordering the Gulf, except Iraq, carried out the most detailed survey ever of the Gulf's physical, chemical, and biological status. Their precise measurements of the Gulf's weak currents will lead to better computer models for predicting the path of spilled oil. Investigating the entire Persian Gulf from Kuwait to the Gulf of Oman, the researchers hoped to determine how the movement of oiled sediment affects the dispersion of oil within the Gulf, especially from the intertidal zone to the subtidal zone. They also investigated the process by which bacteria convert oil into living tissue, looking into both the rate at which it occurs in the Gulf and the question of whether it leads to unforeseen side effects. Particularly important will be their findings on the health of sea grasses and corals, which are fundamental elements of the food chain, and which are believed to have been seriously damaged in certain areas by the spill and months of fallout from the oil-well fires. Led by Robert Clark and Sylvia Earle of the U.S. agency and sponsored by the U.N. International Oceanographic Commission and the Gulf-area Regional Organization for the Protection of the Marine Environment, the cruise was organized and got under way in record time for such a project. Usually five to ten years of planning and diplomatic arrangements precede an oceanographic cruise involving so many countries. It may take that long, however, before the scientists have fully analyzed the data and are able to give a reliable report on this checkup. Even then, they will not be able to address directly the damage caused by the war to the Gulf. Since no such assessment was ever completed before the war, the scientists will find it difficult to compare the findings from the *Mount Mitchell* cruise to the status of the Gulf before Operation Desert Storm.

Despite their importance to the Gulf's ecosystems, coated mangrove thickets and sea-grass beds never symbolized the devastation of the oil spills the way that dying birds did. From the first video images of the spill near al-Khafji to the media coverage of the National Commission's bird rehabilitation center near al-Jubail, it was the agony of the birds that made the world aware of the smothering power of crude oil. Cormorants of both types were the hardiest birds the center treated, and the highest number of them survived, but the number of birds smothered or poisoned to death was many times greater than those rehabilitated. Abuzinada estimated that at least 30,000 seabirds died as a direct result of the spill. But that figure does not include either the deaths among waders and other birds or the thousands that may have suffered severe respiratory damage from the toxic fumes rising from the spilled oil. (The toxicity of oil fumes to wild-life was demonstrated, at the absurd cost of $81,300 per rehabilitated otter, in the wake of the *Exxon Valdez* disaster.) An unknown number of these birds might have had sufficient strength to leave the area, only to die prematurely elsewhere or fail to reproduce. The destruction of habitat and loss of reproductive function will also afflict the one to two million migratory wading birds that pass through the region every year.

It is not known whether any species of bird was brought to extinction by the Gulf War spill, including the threatened socotra cormorant, whose range is restricted to the area. According to the World Conservation Monitoring Centre in Cambridge, England, a group created by the World Conservation Union, the U.N. Environment Programme, and the World Wide Fund for Nature to provide information on biological diversity, the socotra cormorant has been in decline recently as a direct result of "disturbance of breeding colonies and oil pollution." The enormous amount of oil that remains along Saudi Arabia's shores is a serious threat for the cormorant, which forms nesting colonies that can include tens of thousands of birds on sandy shores—the very areas where the greatest quantities of oil are currently found. The continued release of oil from sandy spits and peninsulas further threatens the birds, which typically feed by plunge diving, often en masse, and which could therefore suffer the effects of the Gulf War's environmental legacy for years to come. The loss would be sad, for the sight of hundreds of the birds flying low over the Gulf, suddenly diving, then rising, circling, and plunging again, is a thrilling natural spectacle.

According to Abuzinada, the al-Jubail rescue center had great symbolic value for the people of Saudia Arabia. Others in the country explained that it not only provided an opportunity for those who were eager to stem the horrible tide unleashed by Saddam Hussein but also expanded the environmental consciousness of the entire country; it was a new example of the ways in which Muslims could fulfill their duty as custodians of God's creation.

The Gulf is one of the few remaining habitats of rare marine turtles. Two species, the green and the hawksbill, nest on Gulf islands that were coated with oil; two others, the leatherback and the loggerhead, nest in southern areas of the Gulf and adjacent areas of the Arabian Sea. The World Conservation Union includes all four in its Red List of Threatened Animals, giving the status "endangered" to all except the loggerhead, which is cited as "vulnerable." The World Conservation Monitoring Centre has declared that the population of the Gulf's green turtles that nest on Karan, Kurayn, Jana, Harqus, and Jurayd islands—all of which were heavily oiled—"is of major international importance." Karan Island is their primary nesting site; every spring, 80 percent of the females lay and hatch their eggs in its sandy beaches. Although ecologists have not spent a great deal of time investigating the health of the green and hawksbill populations, they believe that these turtles have been in decline due to oil pollution in the Gulf, particularly the pollution that resulted from the Nowruz and Ardeshir disasters of 1983. The two million barrels of oil released then, the second-largest oil spill in history until the Gulf War, surely harmed the turtle populations: Greenpeace says that the green and loggerhead turtles neither detect nor avoid oil slicks and cannot distinguish between tar balls and food.

Since the breeding season for the green turtles was only a few weeks away when thick tides of weathered crude oil washed up on Karan Island, MEPA placed the island at the top of its prioritized list of sites to be cleaned up, and the International Maritime Organization readily offered to use a portion of its Gulf Pollution Disaster Relief Fund to prevent the threatened environmental tragedy. It hired Alba International Ltd. of Aberdeen, Scotland, to clean up the oiled shoreline in an operation that ran counter to almost all MEPA's principles of keeping additional human disturbances to a minimum. From mid-April to early May, Alba transported its bulldozers on barges across the more than thirty miles of water separating the island from Ras al-Zawr and brought some 19,000 cubic yards of tarred sand from the beaches to pits dug inland, using the fresh sand from the pits to restore the beach contours. According to MEPA administrator Hosni Abdurasek, "We had to take those drastic measures because of the unique importance of the island."

The International Maritime Organization claims that the heavy-duty cleanup was a success, citing follow-up surveys that found encouraging numbers of eggs had hatched. Since the effort was a crash program to salvage the 1991 breeding season, the group's position has some justification. Yet tons of heavily contaminated sand remain on the island, and not just in the disposal pits. "It doesn't take much digging on the beach, only a foot or so, to find sand that's practically dripping with oil," reported Sylvia Earle, chief scientist for the U.S. National Oceanic and Atmospheric Administration, who visited the island in July after the turtles had come and gone. Greenpeace's survey of the Gulf found that many green turtles had skin damage, including some deep skin lesions, but hesitated to link the distress directly to oil. They did, however, find a dead turtle with "substantial amounts of oil in its internal organs," according to Horsman. "This is the only report of such work and indicates both a lack of willingness to do this and a lack of real information to support anyone's statement that the spills caused little or no effect out in the Gulf," Horsman said.

Greenpeace further noted that smoke from the oil fires in Kuwait reduced surface temperatures in the Gulf by 9 to

18 degrees Fahrenheit. Such cooling is significant because sand temperature is related to the sex ratio of turtle hatchlings. At an average temperature of about 82 degrees the ratio is even. Higher temperatures result in more females, lower temperatures in more males. If a disproportionate number of males hatched in 1991, Greenpeace believes, the long-term damage to the Gulf's turtle populations may last for generations.

Neither Greenpeace nor the Saudi agencies have been able to attribute to the oil spill the deaths of about a dozen turtles found on Saudi beaches. There is a similar uncertainty about the roughly one hundred marine mammals that Greenpeace reported had died in the western Gulf between late February and mid-April 1991. The grim list includes fourteen dugongs, fifty-seven bottlenose dolphins, thirteen humpback dolphins, and a number of animals too decomposed to identify. Greenpeace noted that, since most of the carcasses were found south of the areas where the spilled oil had its worst impact, the spills may not have been the direct cause of death. It added, however, that dolphins are particularly vulnerable to oil pollution because, even though they can detect it, they do not appear to avoid oil slicks. Twice during the 1991 survey Greenpeace researchers saw dolphins traveling in oil slicks when clean water was close by. The rare dugong, whose common name in Arabic translates as "the fisherman's wife," reportedly does not avoid swimming in oil sheen either, according to Greenpeace. Like the loggerhead turtle, this distant relative of the manatee is listed as "vulnerable" on the World Conservation Union's Red List.

If the vast contrast in the quantity of oil spilled was the first striking difference between the *Exxon Valdez* and Gulf War disasters, the final difference is the amount of human resources devoted to cleaning oiled shorelines and aiding habitat restoration. Since the crews rolled up their hoses after spraying the mangroves on Gurma Island, the oil on Saudi shores has gone "virtually untouched," according to Sylvia

Earle. The delay is puzzling, since Saudi leaders like Abu-zinada had publicly stated that long-term remedies are essential to reverse the deterioration of coastal ecosystems and restore their viability, and MEPA completed its exhaustive shoreline cleanup plan in July 1991. MEPA projected that a full-scale cleanup and restoration effort would take eighteen months and more than $400 million to complete. Westerners may find it difficult to imagine that Saudi Arabia does not have that amount of money to spend on cleanup and restoration, but by all accounts money is the reason MEPA's plan has been collecting dust, not oil.

The plan is based on the belief that once the oil soaked into the coastal sand and mud, most of the damage had been done, and frantic efforts to reverse that damage were probably going to exacerbate it instead. "We want to assist the cleansing and healing that take place naturally, not presume to improve on it or hasten it unreasonably," MEPA's Abdul-Jaleel al-Ashi explains. He takes as an example a heavily oiled stretch of sandy beach. "The waves and currents brought this sand to this beach with no help from us, and as they did so they sifted the sand so that the size of particles you find here you will not find at the beach a few hundred meters away." If MEPA were simply to take the contaminated sand away, dump it in the desert somewhere, and replace it with desert sand, al-Ashi believes that it would be impossible to match grain sizes precisely and that shortly after the desert sand was brought to the beach, waves and currents would take it away, leaving no beach and therefore no beach habitat. The failure of "beach nourishment" programs in the United States supports this view.

Battelle Ocean Science's Steve LeGore and, significantly, Greenpeace agree. LeGore explains that "aggressive cleanup techniques are often at odds with the integrity of the most productive areas of the intertidal zone, algal mats, for instance." You can't, he points out, get rid of the oil by mechanical means because you would inevitably trample the mat into a foul-smelling, lifeless paste in the process, and Saudi Arabia has always been extremely wary of using chemical dispersants, which could allow the oil to reach cooling-water and desalination-plant intakes. The chemicals could also have unpredictable and damaging effects on the environment in years to come. "There's not much that can be done where oil coated wide areas of tidal flats, except to hope that the algal mat recovers quickly on its own." LeGore sighed as we stood near the edge of such an area on ad-Dafi Bay. And indeed studies of cleanup techniques in Alaska found that recovery often comes more quickly to areas where the efforts were less energetic.

But an International Maritime Organization representative thinks the Alaskan studies may not tell the whole story. "Maybe the barnacles start coming back sooner to the rocks that still have tar on them," he said. "But maybe in a few years we'll see that the recovery on rocks that had been steam cleaned is stronger and goes faster once it gets started."

While waiting for the money necessary to implement the cleanup plans, MEPA has begun developing some new techniques that would keep additional disruption of the environment to a minimum. Most promising is a Rototiller towed behind an ordinary farm tractor, with high-pressure water jets that spray down onto the rotating tiller blades. Particularly effective on sandy shores, the tractor pulls the Rototiller through water about three feet deep, where it stirs up the oiled sand, loosening remarkable amounts of oil from the sand particles. The technique has the advantage of keeping shoreline sediments essentially in place and has led to the recovery of about thirty barrels of oil in a hundred-yard stretch of beach thirty to fifty feet wide. If similar amounts of oil could be collected from the entire length of sand beaches affected by the spill, a thousand more barrels could be removed from the coastal environment.

A thousand barrels of oil does not seem like much in the context of the largest spill in history, but, as Greenpeace points out, "islands" of clean sand or mud can serve as points of recovery that can speed up the process as a whole, although "such activities can be themselves destructive, and careful consideration is needed." Some of the International Maritime Organization's fund has gone to the protection of the few remaining unoiled salt marshes on the Saudi coast, in the hope that plants and animals from the unaffected areas may be used to "seed" the recovery of heavily contaminated areas.

Virtually everyone who has been involved in the cleanup agrees that only time is capable of restoring the coast to a state of ecological stability. Horsman expects that such a state will bear everlasting scars of the Gulf War. In November 1991, LeGore was encouraged by the apparent rate of recovery along the shoreline. He had earlier thought that it would take about a hundred years for the shoreline to recover, but he suggested that ten to twenty years may be closer to the mark. "In moderately oiled areas, the figure may be more like two to five years," he said. "But significant question marks remain—notably, the rate at which the algal mats recover."

CHAPTER EIGHT

Lasting Scars

W H E N   T H E   E M I R
symbolically put out the last of the oil-well fires on November 7, 1991, the people of Kuwait were breathing better than they had in fifteen months. The sky was blue overhead and tinged, on the horizon, with the typical dusty beige of the Arabian desert. Container ships and traditional cargo-carrying dhows plied the mine-free and mostly oil-free waters of the Gulf, restocking Kuwait's stores of everything from caged birds to air-conditioned four-wheel-drive Jeep Cherokees. In parklands along the shores to the east of Kuwait City, once ringed in razor wire and entrenched with Iraqi fortifications, palm trees again provided shade on grassy lawns, from which one could see down to a calm and peaceful sea. Along a wide beach at the northern edge of the city, people crowded into a lively open-air souk, where fresh fish, pottery, kitchen utensils, rugs, and children's toys were bought

and sold to the familiar accompaniment of Arabic bargaining. In the city, the first-class hotels were back in business. In the residential neighborhoods, crews were busy building new houses and repairing those that hadn't been badly damaged during the occupation. At Kuwait International Airport, dozens of young women from the Philippines and other south Asian countries arrived each week, ready to begin careers as domestics in Kuwaiti homes.

Despite all this, one didn't have to look far to see the scars of the Iraqi occupation and the pollution crisis. Back in February, retreating Iraqis had set fire to the luxury hotels where they had established command centers, and the outside walls of these high-rise buildings still bore large sooty stains. The emir's al-Seif palace near downtown Kuwait City was still a shambles from the looting and vandalizing of Iraqi soldiers. Months of fallout from the oil-well fires had left black marks around the roofs and windows of buildings throughout the city. But if you had flown into Kuwait after dark, conducted a day or two of business in town, and left on a flight that departed after nightfall, you would have missed the overwhelming evidence of a pollution disaster comparable to Bhopal's or Chernobyl's.

For that, you had to get out of town or up into the air during daylight hours. From either vantage point the view took your breath away. Oil droplets and soot combined to coat fully a third of the country in an asphaltlike layer that varied from a papery crust to a tarry goo three to four inches deep. Oil lakes and swamps, held tenuously in place by long dikes of hastily bulldozed desert soil, spread across much of the Burgan field in the south and the Sabriya field in the north. Fumes from the lakes stung one's nose at distances of more than half a mile, and for as long as the lakes remain, they will continue to be a deathtrap for migrating birds lured by the marshy appearance of their highly reflective surfaces. No one knows how deep the tens of millions of barrels of oil might be seeping into the desert soil and fractured bed-rock. In the south, and especially off to the west and north, mile after mile of Iraqi trenches rip through the desert in ugly zigzags.

"We will never be able to rebuild the Kuwait we once had—the Iraqis and the oil fires have taken that away from us forever," a man told me. His was a lament any visitor was sure to hear from virtually any Kuwaiti. "We will probably never know how many days of our lives have been taken away by the smoke we had to breathe," he went on. "But, thank God, at least we can live our lives in peace again. As bad as the pollution crisis was, as much rebuilding as we have to do, our hardships today are easy to bear compared to the occupation, when we had to buy the lives of our captured fathers and brothers if we could, or lose them if we could not; when we had to risk our own lives and the lives of families in attempts to escape from occupied Kuwait; and when parents found their newborn babies murdered in the hospital."

During the occupation, the Kuwaitis who remained in their country suffered not only these forms of brutality but also the threat of poisoned water and attack by chemical, biological, and radioactive weapons. The Iraqis were highly suspicious of any sort of resistance action. In the laboratory of Kuwait's Environmental Protection Department, which the Iraqis controlled as they did every other Kuwaiti agency they deemed useful, Kuwaiti technicians routinely tested Iraqi water supplies for poisons that the occupiers believed Kuwaiti saboteurs might use to contaminate the supplies. It was not difficult, according to the department's director, Ibrahim Hadi, for the technicians to slip in samples of their own water and analyze them for the same poisons.

Practically all services vital to public health ceased during the occupation. More than half the population of Kuwait were expatriate workers, who left the country as soon as possible after the invasion. The only businesses the Iraqis kept open, outside of those that served some special purpose for Saddam Hussein, were ones the occupiers considered essential, and their choices were frequently bizarre. Treating sewage was not considered essential. Kuwait's system, which purifies domestic sewage to the point where its outflowing water is fit for irrigating food crops, fell into disrepair. For almost a year after liberation, then, Kuwait had no alternative but to pump untreated sewage into the Gulf. Yet in the first year after occupation, Kuwaitis managed to open their hospitals and specialty clinics, re-equip looted laboratories for research and environmental analysis, and rebuild their systems for monitoring air quality and predicting weather. "It has been a terrific struggle to do so many things in so little time and under such difficult conditions," said Hadi, "but no one in Kuwait would trade all the hard work of rebuilding he did under black skies for living under Iraqi occupation."

The deadliest forms of pollution that threaten people and animals in Kuwait, however, are land mines and unexploded ordnance. For months after the liberation, Kuwaiti newspapers featured an "explosive of the week" on their front pages, complete with a photograph, a description of the munition's power and intended use, and a telephone number for civilians to call to report sightings. In an office at the Ahmadi headquarters of the fire-fighting effort, a grislier photograph was displayed, reminding workers of what their hands might look like if they picked up an unfamiliar object for closer inspection. Within the first year after liberation, more than 1,300 civilians were killed or wounded and fifty bomb-removal specialists lost their lives. The Iraqis planted more than half a million mines in an attempt to seal off Kuwait's coastline and southern border, and many tons of unused ammunition and explosives were abandoned in their bunkers when they left Kuwait. According to Kuwait's Ministry of the Interior, the invading soldiers planted more than 11,000 mines along a single five-mile stretch of shoreline, that is, almost one mine every two feet.

Surprisingly, the Iraqis kept track of nearly all the mines they laid in Kuwait, apparently in anticipation of the time when their control over the country would be recognized and they would be able to administer it as a normal Iraqi province. Presumably they intended at that point to collect the mines. Coalition forces recovered some of the records of Iraqi mine-planting activity and destroyed the mines they found, but no one expects that the last mine will ever be discovered.

And even if every Iraqi mine in Kuwait were destroyed, only a fraction of the threat would be removed. Coalition forces dropped something like 100,000 tons of explosives on Kuwait, a third of which never exploded, according to press reports. The bulk of what remains are the finned bomblets known familiarly in Kuwait as rock eyes or lawn darts, the contents of cluster bombs; each bomb dropped by Coalition aircraft scattered four hundred of them over the Kuwaiti desert. The frighteningly high percentage of unexploded rock eyes is partly a result of the soft desert soil. Had they landed on a harder surface, like the European turf their designers imagined during the Cold War, more of them would have gone off.

Several factors peculiar to the Gulf War and its aftermath complicated the job of the explosive-ordnance-destruction specialists, who work for a handful of companies that contract for dangerous assignments in lands from the Falkland Islands to Cambodia. In and around the oil fields, fallout had turned everything a uniform shade of black, so the crews had no color contrast to assist them as they scanned the desert from the fronts of Land Rovers, slapping the hoods to signal the drivers to stop when they spotted a suspicious-looking item. When they identified a mine or a rock eye, they planted a small flag next to it so that they could come back later and dispose of several at a time. But they had to come back within a day or so, lest windblown sand and dust—the other complicating factor—bury the traces.

If one takes a short walk in the desert outside Kuwait City, the puffs of tan dust that billow up at every step remind one that Kuwait is the dustiest place on earth. Borne aloft by the strong northwestern wind known as the *shamal*, the dust envelopes Kuwait and gives a certain softness to the atmosphere. May and June are the months of the *shamal toz*, the dust-storm season, when the wind howls and the air can become so thick with dust that visibility drops to a few yards.

A short walk in the desert would also reveal why the dust storms are not even more severe. A so-called desert pavement of pebbles and small stones covers the dust in a chaotic but remarkably consistent pattern. According to Egyptian-born geologist Farouk El-Baz, the director of Boston University's Center for Remote Sensing, the stones and pebbles effectively still the winds at the surface and prevent the microscopic dust particles beneath them from being lifted up into the air. "Even though the wind can easily carry the desert dust away if there is no layer of pebble and stones on top," El-Baz explains, "that layer protects the dust particles and holds them in place; if you dig with your shovel and remove the pebbles and stones, you increase by many times over the number of particles that the wind can carry away." Digging also disrupts the delicately sifted, packed arrangement of the microscopic particles. For the desert soil to remain in place, the packing of the tiny particles is as important as their stony armor. Walking across the undisturbed desert may raise puffs of dust at every step, but walking across a disturbed area, such as tank tracks, is an experience of a different order: one can sink in past one's ankles and quickly raise a choking cloud of dust.

Centuries of nomadic grazing and decades of oil development have done little to damage Kuwait's desert pavement, which, by preventing massive dislocations of the dusty
soil under it, also allows a sparse selection of shrubs, grasses,
and sedge to thrive. Typically, the plants grow within the
vague confines of wadis, where they can make the most of
seasonal showers. Since those usually dry riverbeds are the
only places where material carried by intermittent streams
can collect, they contain Kuwait's most fertile soil. Annual
plants survive there by telescoping their life cycles into the
brief periods when sufficient moisture is available to pro-
mote germination. They usually have short stems and roots
and, while in flower, create a fragrant but wispy carpet over
the desert surface. Perennial plants survive by developing far-
flung root systems to capture the available water over a wide
area and support the above-ground growth, which covers a
much smaller area; hence the spacing between plants, which
echoes the speckled appearance of the desert floor. An assortment of reptiles, birds, and rodents are also able to sustain themselves in this environment, living for the most part
in burrows or under rocks. The delicately balanced ecosystem depends entirely on the stability on the dusty topsoil.
Changes in the distribution of the soil or in its permeability
to water could destroy it altogether.

Shortly after the Coalition forces launched Operation
Desert Storm, El-Baz issued a warning that the military activities of both sides would produce wide-scale disruption of
the desert pavement in Kuwait and southern Iraq, leading to
increasingly frequent and severe dust storms. Events have
proved him right. A group of Kuwaiti scientists who remained in the country during the occupation and banded
together to make the initial assessments of the postwar environment found that all the armies, but especially the Iraqis,
imposed "terrific disturbance on the surface of the desert."
The group, which called itself the Kuwait Environmental

Action Team, went out in the field even before the last skirmishes and discovered an extensive system of trenches and bunkers in western and northern Kuwait. They were shocked by the enormous amount of soil the soldiers had dug up that now lay exposed and vulnerable to the wind. The "end run" through southeastern Iraq by armored divisions of Coalition forces did a comparable amount of damage to the fragile desert pavement. Speeding across the desert and raising huge clouds of dust in the process, the tanks left behind a wide swath of unconsolidated desert soil.

Within two weeks after Iraq agreed to the terms of the cease-fire in late February, "the howling winds that blow across the desert" were spraying "a sticky grit over everything," according to a report in the *Washington Post*. El-Baz traveled to Kuwait in May and saw that his worst fears had been realized. He warned General P. X. Kelley, commander of the U.S. Marine Corps in Kuwait, that the large areas of disturbed soil would lead to the creation of sand dunes that could grow and move fast enough to obscure roads and aircraft runways overnight. The drifting sand was also beginning to cover areas where large numbers of mines and other munitions remained unexploded.

General Kelley agreed to do what was possible with the limited number of troops still in Kuwait at that time. But by the end of the dust-storm season, El-Baz could say that "the *shamal toz* of May and June 1991 was one of the worst ever." He had earlier seen how battles in the 1973 Yom Kippur War led to changes in the configuration of sand dunes in the Sinai Peninsula, and he knew that the number of dust storms in the Gulf region doubled in the wake of the Iran-Iraq War, often forcing airport officials to close their facilities. In the months following the Gulf War, the dust storms in Kuwait were wilder and occurred faster than El-Baz had expected, and in their sudden ferocity, they nearly filled in all the Iraqi trenches and bunkers with dust from the northwest. "I really thought that it would take several years for the situation to develop to this point," El-Baz said. Buried was a vast quantity of ammunition and other armaments, but much of the soil the Iraqis had dug up still lay in piles beside the filled fortifications. "This means that the increased severity and frequency of dust storms will not stop but, rather, continue for several years," El-Baz explained.

If the dust storms continue to rage with greater intensity and frequency in years to come, as El-Baz believes they will, the danger of accidental explosions in northern Kuwait may actually rise rather than fall, despite the efforts at mine sweeping. *Shamal* winds could swiftly bury thousands of tons of lethal ordnance in depths of sand sufficient to hide them from detection instruments. Fortunately, they probably also will be sufficiently deeply buried to pose little harm to animals or people if they detonate. Indeed, one strategy for mine-proofing an area consists of bringing in tons of fresh soil and packing it down tightly. But in the desert, where explosive-ordnance-destruction crews destroy all the munitions they find and then move on, dusty sand doesn't remain on the spot where the wind dropped it. In five, ten, or fifty years, once-buried lawn darts that were beyond the reach of vintage-1991 detection equipment may again emerge, with all their explosive power intact. "The increased activity of the dust storms is going to make it a real problem to figure out where the explosives are," El-Baz says. "A lot more of Kuwait is going to be off limits to people and animals than I ever thought would be necessary, and it will be a very long time before the desert is safe again." According to El-Baz, mines planted during World War II in the western desert of Egypt still explode and cause casualties.

El-Baz laments the lack of a regional system to monitor or predict dust-storm activity, which could be used to identify areas where the wind is depositing the largest amounts of material. Kuwaiti officials could then have explosive-ordnance-destruction crews clear those areas first, before it became too difficult and time-consuming. Although there are well-equipped weather stations throughout the Gulf region and these stations provide reliable information on the areas in which they are situated, the fact that they are not tied together in a regional network leaves the environmental-monitoring capability in and around the Gulf sorely inadequate, in El-Baz's opinion. He says that "nobody has the slightest notion of what is going on with the transport of particles on the surface of the desert," and "regional forecasts of wind direction and speed are necessary to alert officials to the possibility of dust storms."

While Kuwaitis will continue to face the threat of mines for decades, their descendants will find evidence of the Gulf War's environmental consequences for centuries, if not millennia, in the quantities of liquid that gushed unburned from out-of-control wells and fell as tiny droplets from the sky. The amount of oil that soaked into the earth from January to November 1991 will never be known with any precision, but it was certainly hundreds of millions of barrels, comparable to the total amount of motor gasoline burned in California during 1989. Throughout the well-fire crisis, the world fixed its attention on the incredible amounts of smoke the fires pumped into the atmosphere, but most of the people who witnessed these fires firsthand say that the flames consumed only about a third to a half of the oil erupting from the ground. If the Kuwait Oil Company was correct in estimating that 3 percent of the nearly 100 billion barrels of proven reserves was lost in the catastrophe, then at least 1½ billion barrels either spilled directly onto Kuwaiti soil or rained down on Kuwait, Saudi Arabia, Iran, Iraq, and the Gulf. This is more than 130 times the largest estimate of oil spilled from tankers and export terminals into the Gulf as a result of the war and nearly 6,000 times the amount spilled in the wreck of the *Exxon Valdez*.

As much as 30 percent of the unburned oil—a relatively high percentage—could have evaporated, considerably reducing the quantity remaining in the soil, in oil lakes, and in the Gulf. But what is one left with if one reduces an unimaginable quantity by 30 percent?

As the oil lakes in Kuwait formed and grew during 1991, they followed the paths of least resistance, into the wadis in and near the oil fields. Once in those natural drainage systems, the oil saturated the most fertile soil in the country. The lack of any permanent rivers or streams spared Kuwait's neighbors and the Gulf from having to deal with any more oil than they were already faced with, but the extreme flatness of the terrain allowed the lakes to stay shallow but cover more area.

Estimating the quantity of oil in the lakes, and the area covered by them, is difficult for several reasons. Spot measurements of the depth to which the oil penetrated under the lakes have failed to reveal a consistent pattern, so there is no reliable way to determine how much oil saturates the ground under the lakes. Speaking at a conference on the environmental effects of the well blowouts, Adnan Akbar, director of the environment division of the Kuwait Institute of Scientific Research, noted that if the oil reaches the bedrock underlying the desert surface, it could penetrate deep into the earth because the geological formations of the region are riddled with fractures. As it stands, the lakes drift off sluggishly in response to every change in the wind, assuming new configurations and saturating new areas.

Amazingly, some young plants have emerged from soil soaked inches deep with oil, and within weeks of the killing of the last wild well, fresh lizard burrows began appearing in the oil-drenched sand of the Burgan field. But with such unprecedented contamination, environmental scientists can only guess at how severe, widespread, and long-lasting the effects of the oil lakes will be. It is clear that the sooner the standing oil is removed, the less impact it will have. Although the Kuwait Oil Company began recovering oil from the lakes even as the fire-fighting effort was coming to a close, its effort was driven more by financial than by environmental concerns: the oil in the lakes contains much of the salt water that rose through the wells with it, but it can be refined, and the company plans either to do so itself or sell the recovered oil to another refiner. When it decides that the cost of recovering the standing oil is greater than the environmental and monetary benefit, the effort will end, and wind-blown dust will slowly cover the oil still remaining on the desert. Perhaps geologists of the future will marvel at the thick, dark band that dates back to sometime near the year 2000.

❦ ❦ ❦

HIGH ABOVE THE oil lakes, obscuring the sun and cooling the desert surface, the smoke plume from the oil-well fires hung over Kuwait like a gigantic ethereal demon for most of 1991. By the time it rose a thousand feet or so, one could no longer distinguish the contribution of individual fires. Generally, there were two composite plumes, one from the northern fields and the other from the fields to the south of Kuwait City. Merged, they formed the so-called supercomposite plume, which gradually dispersed as it drifted away, typically to the east and southeast. As it moved through the lower atmosphere, it often split into two or more tiers, according to the distribution of stable layers in the air. Seen from above, it assumed every conceivable shape, sometimes smooth, sometimes wavy or puffy; seen from below, the perspective of the average Kuwaiti, it was usually flat, like a ceiling, and black.

The first reports of the incredible sight were usually accompanied by expressions of fear that the high content of sulfur in Kuwaiti crude would give rise to massive amounts of airborne sulfur dioxide—a gas that can be deadly if inhaled and that can combine with water droplets in the air to form acid rain—and no one had any doubt that the smoke should be considered a grave, if ill-defined, threat to the health of people and animals. So the authorities in postliberation Kuwait gave a high priority to reestablishing the disrupted network of weather-reporting and pollution-monitoring stations, and air-pollution specialists from around the world flew to Kuwait as soon as possible. The reports, in March, of autopsies on dead sheep that showed lungs spotted with black blotches heightened the level of anxiety. Within a month of the liberation, the U.S. Public Health Service and the National Oceanic and Atmospheric Administration had assisted Kuwait in rebuilding its system of monitoring towers to report and predict the plume's trajectory.

The smoke contained far less sulfur dioxide than almost anyone had anticipated. Two scientists from the Kuwait Environmental Action Team, Ali Khuraibet and Fatima Abdali, went into the field during the first weeks of the crisis and took air samples from as close as sixty-five feet from the fires. They detected neither sulfur dioxide nor the similarly dangerous class of gases, the nitrogen oxides. As scientists from other countries arrived with increasingly sophisticated equipment and sampling techniques, Khuraibet and Abdali's findings were confirmed, and a lively debate developed about where the sulfur dioxide was "going." By the end of 1991, essentially all the scientists engaged in the discussion had come to realize that within the plume a puzzling array of chemical reactions was rearranging the molecular products of the oil fires. Many of them came to believe that the high concentration of salt in the plume was somehow reacting with the sulfur, causing its transformation into chemicals they were not equipped to measure at the time, and consequently decreasing the acidity of both the plume itself and the rain that fell through it.

The air quality in Kuwait City was horrible, according to many residents, and in the town of Ahmadi, amid the southern oil fields, it was far worse. When Kuwaiti men in either town went outdoors for just five or ten minutes, their white ankle-length shirts, or *dishdashes,* invariably become speckled with black droplets of oil that had fallen from the sky. In Ahmadi, the black spots were a little larger and spaced a little closer together. But bad as it was, it could have been much worse. The heat of the fires quickly boosted the smoke up to altitudes of several hundred feet, and although the heavier oil droplets and bits of soot rained out of the plume, the plume itself usually remained high above the people as they went about getting their lives and country back together again.

The efficiency with which the plume blocked sunlight usually contributed to keeping the soot high aloft, since with no surface heating by the sun's rays, there was no rising air to mix with sooty air above and bring it down. Sometimes at sunset a meteorological phenomenon known as a thermal inversion would occur and hold the plume close to the ground. At such times there was no relief except to stay indoors, where the air was generally better but still noticeably poor, and wait until sunrise, when the inversions would break. More than any other thing, the fact that the worst of the pollution was usually overhead and not at ground level protected the residents of Kuwait from its most threatening effects.

On occasion, the concentration of sulfur dioxide in Kuwait City's air reached levels recognized as dangerous by the U.N.'s World Health Organization. These were often short-lived episodes, but a team of air-pollution scientists from France who took measurements during March and April found sulfur dioxide concentrations in Kuwait City that exceeded the organization's guidelines. More persistent than any chemical, however, were the inhalable particulates, but even these did not necessarily endanger people and animals. After all, since Kuwait is always extremely dusty, public-health officials did not know, even when particulate readings soared, what fraction of the material in the air was the dust that everyone had been breathing since birth and what was oil-fire soot. Even though the soot did not seem to be as dangerous as many thought it would be, no one was in a position to say that it was benign.

Faced with an unprecedented air-pollution disaster but possessing information that showed that the pollution might not be as harmful as feared, Kuwaiti officials could not decide whether to encourage the Kuwaitis who had fled the occupation to come back and begin rebuilding their country or to play it safe and emphasize that the ultimate danger posed by the fires remained a mystery. By April, they had found a middle course and said that the air in Kuwait City was not as dangerous as it might have been but that citizens abroad who had known respiratory problems would do best to remain where they were until all the fires were out. At about this time, a controversy arose as to whether Kuwait City's air was actually polluted at all. A number of Kuwait Oil Company employees who endured long periods of heavy smoke close to the ground in Ahmadi said that "the smoke never got to Kuwait City" and "the people of Kuwait City were never affected by the smoke."

People who spent the summer in Kuwait City, however, could only shake their heads at such statements. A secretary in the Ministry of Information told me that outside her home, about two miles south of the center of town, she could swish the soot in the air with her hand on many days. According to Ali Khuraibet, nurses in the intensive-care ward of Kuwait City's maternity hospital had to change the air filters on the babies' incubators about once every two days during the worst soot episodes. Previously, they had changed the

filters about once every six months. Tony Horwitz, a journalist who covered the Gulf War and its aftermath for the
Wall Street Journal, said that "the air pollution was one of
the most amazing things" he had ever seen. "When we started
hearing official statements that it wasn't so dangerous," he
said, "it really went against the evidence of the eyes."

CHAPTER NINE

Breathing Black Air

W E W I L L P R O B A-
bly never know precisely how dangerous Kuwait's air was
to breathe in 1991—just as we may never know definitely
whether smoke from the oil fires played a role in the Asian
floods that killed thousands late that same year. For all the
air sampling that Kuwaiti, American, and other scientists
carried out, for all the concern international environmental
agencies and organizations focused on the soot, and for all
the anxiety residents of Kuwait and nearby areas felt as they
gazed up, day after day, into blackened skies, the officials
ostensibly responsible for answering the key question—what
is going to happen to me and my family if we breathe this
air?—concentrated on less important questions. In their public
statements, they skirted the real question, and they failed to
undertake, commission, or even allow studies that could have
answered it. Instead, they withheld pertinent information from

scientific scrutiny. Public-health researchers both in and outside Kuwait have given up because Kuwaiti officials have apparently decided to treat as state secrets data from the few long-term studies that tracked the health of Kuwaitis exposed to the well-fire smoke—studies initiated, as it happens, only during the final weeks of the fire-fighting effort.

The fire fighters, of course, succeeded extraordinarily well in taking the action most certain to reduce the public-health risk: putting out the fires. Once the fires were out, air quality improved dramatically, although fumes from the oil lakes still pose a threat of undetermined dimensions. By putting out as many fires as they did by the end of the summer, the fire fighters also averted a crisis in the autumn, when the strong *shamal* wind out of the northwest gave way to calmer air. If many fires had still been burning between September and December, after the *shamal* had died away and could no longer sweep the sooty air out of Kuwait, the smoke would have concentrated in areas near the oil fields and caused the worst air pollution of all. It was not, however, Kuwait's concern for the environment or for the health of its residents but, rather, the value of the oil gushing uncontrolled out of the ground that provided the greatest impetus for swift completion of the capping. At about eleven million barrels a day, Kuwait lost an average of $2,500 a second while the fires burned. Ending the environmental and public-health crises was at best a side effect.

The scientific effort to probe the complexity and magnitude of the fires, carried out both within Kuwait and around the world, was almost as impressive as the achievement of the fire fighters. The data Peter Hobbs and his colleagues from the University of Washington collected as he flew through the smoke plume in May and June, which revealed that the plume rapidly lost much of the sulfur dioxide given off by the fires, demonstrated how incomplete our understanding is of how the gases and particles produced by large fires react in the lower atmosphere. Robert Stevens, a scientist with the U.S. Environmental Protection Agency, who took plume samples from individual fires during low-altitude helicopter flights and analyzed them for "practically every element in the periodic table," complemented Hobbs's work at higher altitudes and farther downwind. Stevens also found the Kuwait oil fires' "signature"—salt crystals so tiny that a thousand of them lined up edge to edge would equal the width of a human hair. The salt that gushed from Kuwait's oil wells as a component of the water that rose with the oil, boiled away to become a gas and then cooled and condensed to form the crystals. Atmospheric scientists can now look for these minuscule crystals to help them trace the progress of the oil fires' smoke in the atmosphere.

Much of the information collected was reassuring for human beings. Low levels of sulfur dioxide meant that rain or snow falling through the plume would be less acidic than feared and that the smoke posed a less acute health risk to those who breathed the air. Several spot surveys of Kuwait's air during the crisis also revealed that certain chemicals linked to cancer were not present in alarming amounts. And the fact that the densest parts of the plume were often more than half a mile up, and usually not directly over Kuwait City, made life less difficult than it might have been. As the scientists amassed their data, officials in Kuwait and the United States eagerly emphasized each new item that might allay the fears for Kuwait's environment. When committees of the U.S. Senate and House of Representatives asked officials of the Bush administration to report on the status of the fires and the possible bad effects of breathing Kuwait's air, it was invariably this heartening information that was provided.

But many scientists, like John S. Evans of Harvard University's School of Public Health, who chaired a conference on the fires in August 1991, found insufficient information among the volumes of data to convince them that Kuwaiti residents were not at risk of suffering serious effects, including death, during times of the heaviest smoke concentration in populated areas. The problem was that the scientists who took most of the air samples in Kuwait during the crisis analyzed them for only a small number of the chemicals that the fires released into the atmosphere. The chemicals they focused on were, for the most part, dangerous, but the analysis of the dangers were based on controlled experiments in which the chemicals were in a highly purified state. In Kuwait, with the fires spewing out millions of different kinds of chemicals, the mix changed from day to day, according to how long the chemicals in the plume had been reacting with one another. Some of the chemicals were gases that, under normal circumstances, would not be as dangerous as they were in Kuwait, where they attached to tiny smoke particles, enabling them to lodge deep within the lungs. Also hindering accurate assessment of the threat was the fact that, during the months following liberation, Kuwait's population was constantly in flux: many who fled the occupation were returning, and people were moving from one area to another.

To get a clearer idea of the acute risks in Kuwait, a few scientists performed toxological studies, exposing test organisms to varying concentrations of unrefined air samples. In one such study, researchers from the Harvard School of Public Health found that pollutants in the air they sampled in Ahmadi during April and May were about as toxic as pollutants found in the air of U.S. cities. While these findings seemed, at first glance, to allay some of the fears for Kuwait's residents, Evans noted that Kuwait's residents were exposed to much higher concentrations of the irritants than residents of U.S. cities were. If Kuwait's environmental and public-health officials had authorized a systematic program of toxicological studies, taking samples at regular intervals at a specified set of locations, a reliable picture of Kuwait's airborne peril might have been pieced together. As it was, each toxicological study conducted in the country during 1991 was an independent affair, and only ineffectual steps were taken to inform the population of the studies' significance.

There was, however, one group of Kuwaiti scientists who tried to establish a systematic approach to environmental and public-health monitoring in the weeks immediately following the liberation. The Kuwait Environmental Action Team (KEAT), organized during the Iraqi occupation, carried out the first assessments of Kuwait's postwar environment. Led by Jassem al-Hassan, head of the biochemistry department at the University of Kuwait, the scientists were mostly young Kuwaitis who earned their Ph.D.'s at Western universities. Between August 1990 and February 1991, they worked to obtain information about the dangers of Iraq's nonconventional weapons and educate the population on precautionary measures. After obtaining Iraqi communiqués detailing the warning signals that would precede attack by chemical, biological, or radioactive agents, KEAT members printed out instructional pamphlets on hidden copying machines. Clandestinely distributed in Kuwait's mosques during prayer times, these told Kuwaitis how to protect themselves in various ways, ranging from keeping a pet bird as a "coal-miner's canary" to creating a kind of fallout shelter within one's home.

Immediately after the liberation, KEAT went into the field and located a number of oil-filled trenches and pipelines spilling oil. In the first critical weeks after the rout of Saddam Hussein's army, KEAT was virtually the only source of information on Kuwait's environment for scientists and journalists. During April, it surveyed about 1,750 people in eight residential areas, five near the Burgan oil field and three in Kuwait City. It found that more than half the residents experienced breathing difficulties, and 95 percent had had their property damaged by smoke and oil droplets from the blazing wells. The survey included a question on whether the government should broadcast information on actions that could be taken to safeguard health and property. Eighty percent of the respondents felt it should.

By May, KEAT was planning a comprehensive study of hospital records to determine more precisely how Kuwait's polluted air was affecting the health of the population. It also began calling for a comprehensive assessment of the environmental damages resulting from the Gulf War. This was about the time that Kuwait's official environmental establishment regrouped in the country, and from that point on, KEAT found its voice diminishing. It had been urging the government to authorize a large-scale program of environmental and public-health monitoring, to be designed and begun at once. KEAT also wanted the details of the program's design and findings to be open for scrutiny by any qualified scientist. Kuwait's Environmental Protection Council, an organization with ties to the Ministry of Public Health, met KEAT's call with the creation of its own studies, designed with help from the U.N.'s World Health Organization. The Ministry of Public Health plans to publish technical reports on the results of these studies, but since the reports will not be subjected to the type of review that usually attends the publication of scientific information, the methods and the conclusions of the studies will remain open to question.

According to KEAT's Sami al-Yakoob, because Kuwait did not follow through on his group's recommendations, the most important questions about the health effects of the oil-well fires may never be definitively answered. "We don't know what the pattern of exposure in the Kuwaiti population was; we don't know what percentage of the population was exposed; we don't know what the health status of the exposed population is, and we don't know who are the people at risk," he explained.

Mamoud Abdulrahim, deputy director of Kuwait's Ministry of Public Health, says that KEAT's appeal was too emotional, bordering at times on the hysterical. He cites as an example the group's recommendation that Kuwait purchase a million gas masks as a precaution against the possibility of deadly air pollution. But Abdulrahim acknowledges the contribution that KEAT made during the first weeks of liberation: "The zeal and energy they showed put the issue in focus."

KEAT's willingness to go to the Western media and propose sweeping environmental programs that would be open to peer review was a stark contrast, and a challenge, to the status quo at the Ministry of Public Health. When KEAT members publicly criticized and contradicted Kuwait's environmental establishment, they stepped outside the norms of traditional Arabian behavior: in the Middle East, disputing parties that consider themselves members of the same "family" rarely air their differences in public, relying instead on personal trust. KEAT also could be seen as taking advantage of the emir's absence, and the consequent power vacuum, during the occupation. Although, for example, there is always a certain amount of tension between Kuwait's ruling minority of Sunni Muslims and the majority of Shiite Kuwaitis, such grievances as the Shiites may have can usually be deferred; they seldom erupt in the light of day. Kuwaiti citizens enjoy a per-capita income among the highest in the world and are free of material cares; there is no unemployment, health care is free, and most manual and bureaucratic labor is done by third-country nationals, who outnumber Kuwaiti citizens. Many Kuwaitis are less than enthralled by the prospect of democracy in their country, lest the lifting of restrictions on the press and the right to assemble be accompanied by a decrease in the government's material largesse.

With its traditional and well-oiled way of getting things done, its lack of environmental consciousness, and its indifference to democratic rights for legal residents, Kuwait appears extraordinary to a Western observer. The country teems with enormous cars that burn leaded gas, and the network of streets in Kuwait City is like a fantasy of the California highway planners of the 1950s and 1960s. Virtually the only people who walk or use public transportation in the city are the poorer third-country nationals and the stateless Arabs known as Bedu or Bedoon. As the fire-fighting effort came to a close, airborne lead emissions from automobiles became one of the pollutants of greatest concern in Kuwait City, according to Robert Stevens of the EPA. For a few weeks after the liberation, water was exceedingly scarce in Kuwait, but gasoline was free. With oil droplets raining down over everything, Kuwaitis who wanted to drive clean cars discovered a new use for the free gasoline: it was a wonderful solvent for the petroleum precipitation that stained auto bodies and obscured windshields. Gasoline car-wash operations quickly became a thriving business for third-country nationals, who spent their days absorbing lead, benzene, and other toxic or cancer-causing components of gasoline. So far, no plans have been announced to examine these residents for the health effects they may have suffered or might suffer as a result of their prolonged exposure.

Understandably, but nevertheless, incredibly, Kuwaiti officials do not want their country to be known as an environmental disaster. When they began trickling back into the country during April and May, they saw the work of their lives in ruins. Not surprisingly, their outlook was colored by a certain degree of denial, and they consistently erred on the side of overoptimism about the threat of air pollution. In the interest of quelling an incipient panic supposedly sparked by KEAT, officials denied any evidence of danger when asked about the risk posed by the oil fires. On one occasion, for example, a question as to whether autopsies had been performed on sheep in an attempt to assess lung damage was dismissed as irrelevant because sheep lungs are not a food item. When KEAT said that governmental inaction and denial were preventing an accurate determination of the air-pollution risk, the Ministry of Public Health replied that they did not want to frighten the population by crying wolf. General interest in the dispute between the two camps did not quite eclipse concern over the issue they were disputing, but the dispute confused the issue for the average Kuwaiti, whose handkerchief was stained black from coughing and sneezing.

"The whole environmental situation in Kuwait has been politicized," said Jassem al-Hassan of KEAT and the University of Kuwait. He was dismayed by scientists from the West who sent technicians to Kuwait for environmental fieldwork on pollution. According to al-Hassan, too much of their research was geared toward the production of papers documenting the disaster for obscure scientific publications, and not enough toward answering the question on the minds of thousands of Kuwaitis. He noted that officials in Kuwait and their counterparts in the United States and institutions such as the World Health Organization trumpeted every finding that hinted at the possibility of minimal risk to the public health, while delaying, or finding excuses for not initiating, studies that might have had greater significance for public welfare. "This led to missing a golden opportunity for doing a reasonable assessment of the fires' atmospheric chemistry, as well as for providing reliable information to Kuwaitis on the health risk posed by the fires," al-Hassan lamented.

KEAT's Sami al-Yakoob of the Kuwait Institute of Scientific Research was particularly upset about what he saw as a failure on the part of the Ministry of Public Health to provide the Kuwait population with an adequate risk-awareness program. "They went from ignorance about the air-pollution risk to saying to the public that the air was safe to breathe," he asserted. In a kind of parable, he asked: "If you are locked in a room with a sleeping lion, can you say that you are safe? You don't know when he is going to wake up or how hungry he might be when he does. If, in this smoke and oil containing thousands and thousands of chemicals, I find that the concentration of five dangerous chemicals is low, can I say that I am safe?"

Journalists, too, reported an apparent reluctance on the part of Kuwait's establishment to face the question of how the polluted air might affect the population. Writing in *The New York Times*, Matthew Wald reported that "Kuwait has not put much emphasis on reestablishing a testing laboratory plundered by the Iraqis that could measure what was going into the air, and had declined to pay EPA to do the job."

There is conflicting information about the question of whether Kuwait's population was adequately apprised of the severity of the pollution and ways to minimize their exposure. By June, Kuwait had established a health-alert system that informed the population whenever any of the five most dangerous pollutants reached the danger level. Almost invariably, inhalable particulates were the only category that exhibited high levels, often because of sand or dust storms. According to the Environmental Protection Department's Ibrahim Hadi, Kuwait's schools displayed posters instructing children to avoid outside play when the plume was overhead and to keep the air inside their homes as clean as possible by encouraging their parents to cook less and smoke less. But when al-Hassan asked his school-age daughter if she had ever seen such posters, she said she had not.

Hadi claims that because Kuwait put its greatest effort into the most effective means of safeguarding public health—putting out the fires as quickly as possible—other worthy efforts were necessarily shortchanged. Kuwait had to do its best with what was left. The health-alert system consisted of a network of air-quality monitors that fed data via phone link to the department, where the levels of the five pollutants were compared to Kuwait's standards. Emphasizing the department's concern to avoid panic among the population, Hadi observed that "if the pollutants detected were above the standards, then we would alert the hospitals to expect an influx of patients." The population was not advised to use air purifiers in their houses or to wear masks when the quality of the air seemed bad. "If the government had recommended such actions," Hadi said, "the prices of those items would have skyrocketed."

Third-country nationals in professional positions or in the employ of the Kuwait Oil Company reflected the eagerness of official Kuwait to downplay the possible health risk. Most of this group works on limited-term contracts, which Kuwait can decide to revoke or not renew for almost any reason. They make a good living in Kuwait, usually far beyond what would be possible in their own countries. For example, Paul King, who organized the effort to keep the fire fighters supplied and fed, said that concern over the air-pollution hazard was overblown, and he offered the opinion that the oil soaking into hundreds of square miles of Kuwait desert might prove a wonderful fertilizer and actually help the country's small but important agricultural industry. A month after the last fire was out, however, the yard around his house in Ahmadi was remarkably free of well-fire fallout, and his family had yet to return to Kuwait from the United States. Gazi Belushi, a public-information officer for the oil company, agreed with King's assessment of the polluted air. But once the fire-fighting effort was all but complete, he wasted no time traveling to Europe for a week's worth of medical tests. Everything was fine, thank God, he said.

Robert Stevens of the EPA feels Kuwait's Environmental Protection Department "did a magnificent job measuring air quality on a daily basis" and diligently kept the population aware of when winds, or lack of wind, might bring smoke into residential areas. "I read the reports as they were given to the radio and to the community, so I know that the department kept the community at large informed about what was going on," he says.

It is difficult to reconcile such statements with those made by members of KEAT, who unanimously agree that, throughout the crisis, government officials consistently reassured the population that the air in Kuwait City was safe to breathe. "They refused to tell the people that the air might be dangerous, because they said that, if they did, people would panic and not work to rebuild the country," Sami al-Yakoob explained to me. "This point of view is ridiculous," he continued, his voice rising. "They were the ones who panicked and fled when the Iraqis came. If the Kuwaitis who stayed put and kept the country alive through the occupation were brave enough to fight against Iraqi soldiers who were armed with automatic weapons, do you think they would not do their duty to rebuild the country just because the air was full of smoke?"

Regardless of how effectively the Environmental Protection Department determined the pollution's health threat or how aggressively it communicated its findings to the population, the message most residents seem to have heard was that the air was not dangerous. But according to most of the Kuwaitis with whom I talked, people relied on the evidence of their eyes and throats rather than on government bulletins. In the Ministry of Information, of all places, a secretary told me that it was difficult to believe the government when it said that the air pollution was not much of a threat. "How could you believe it," she asked, "when you had those scientists from the university and the Institute of Scientific Research saying that it might be dangerous, and when you looked all around you and saw that the air was black?" She shook her head at the bitter division the controversy created in Kuwaiti society. "It's too bad people became so angry with each other about this, especially when we should all have been pulling together, like we did during the Iraqi occupation. It was wonderful how serious we were about taking care of each other then."

Although the Iraqis either looted or vandalized practically all of Kuwait's environmental monitoring equipment and laboratories, within days of the country's liberation the Federal Air Team, a group of environmental experts from the United States appointed by President Bush, had arrived and determined that Kuwait's meteorological stations needed only new power supplies and recalibration to resume recording information on wind and speed and strength. The repairs were completed in a few weeks, according to John H. Robinson, director of the Arabian Gulf Program office at the U.S. National Oceanic and Atmospheric Administration, and the fifteen stations began sending Kuwait's environmental officials information that could be used to predict the path of the smoke plume and to warn fire fighters and troops of when the smoke might interfere with their activities. The team's first measurements of acidic gases such as sulfur dioxide, nitrogen dioxide, and deadly hydrogen sulfide revealed that Kuwait's air usually contained less of these gases than are generally considered dangerous—a most surprising finding, given the high sulfur content of Kuwait's crude oil.

In late March and early April, a organization from Paris, called AIRPARIF, traveled in a mobile air-quality laboratory to eight locations in Kuwait and found that sulfur dioxide levels were, on the average, less than half of what the U.S. National Ambient Air Quality Standards considers dangerous. According to this system, dangerous levels are considered to pose a significant threat "to the most sensitive population, with an adequate margin of safety." It also discovered, however, that over short periods of time sulfur dioxide sometimes exceeded the levels that, maintained over a period of twenty-four hours, are deemed dangerous. Over an eight-hour period on April 3, for example, the French found an average of 495 micrograms of sulfur dioxide per cubic meter of air at Kuwait International Airport. The twenty-four-hour standard for the United States is 365 micrograms per cubic meter. On the night of April 3–4, when the French team's van was engulfed by a smoke plume that touched down in Kuwait City, measurements of sulfur dioxide reached a peak of 2,911 micrograms per cubic meter over a five-minute period and averaged 472 micrograms per cubic meter over twelve hours.

In contrast to sulfur dioxide and the other so-called priority pollutants—carbon monoxide, nitrogen oxides, and ozone—levels of inhaled particulates were consistently high, often in the range that the U.S. standards define as dangerous. The U.S. standard for the category of particulates known as PM_{10}—meaning particles with diameters of ten microns or less—is an average 150 micrograms per cubic meter over a twenty-four-hour period. For practically the entire month of May, three monitoring stations set up by Kuwait recorded averages more than three times the U.S. standard and also above Saudi Arabia's standard of 340 micrograms per cubic meter.

Saudi Arabia's relatively high standard for PM_{10} hints at an ambiguity in measurements taken in Kuwait. In this dusty region, the measurement recorded not necessarily smoke pollution but rather the total amount of inhalable particles, either smoke or sandy dust, in the air. If Saudi Arabia had taken the U.S. standard, its air would have been dangerously polluted practically every day, going back to before Islam. When researchers began analyzing PM_{10} samples for carbon content, an indication of the fires' contribution to the sample, they found that the percentage of carbon varied from 1 percent to more than 50 percent, depending on where the sample was taken and the strength and direction of the wind.

In his testimony to the Senate task force, John Evans gave two examples of how the ambiguity could lead researchers astray. On May 2 in al-Jahra, a town just west of Kuwait City, the PM_{10} concentration was more than 800 micrograms per cubic meter, but less than 1 percent of the sample was carbon, "suggesting that most of the aerosol was not due to the fires." On May 7 in Ahmadi, the PM_{10} concentration was 450 micrograms per cubic meter, but almost 60 percent of the sample was carbon, "indicating that virtually all of the aerosol was due to the fires." Thus, a high PM_{10} reading taken in a residential area might not, by itself, have warranted alerting the population.

When the Gulf Pollution Task Force of the U.S. Senate Committee on Environment and Public Works held its first hearing on April 11, 1991, Federal Air Team leader Jim Makris of the EPA testified that the team found dangerous chemicals adhering to smoke particles. These chemicals included sulfur dioxide, which poses a short-term—or acute—threat to health, and polycyclic aromatic hydrocarbons, which are believed to cause cancer and so present a chronic threat. But "the team did not detect such chemicals in any significant quantity," according to Makris. Even so, Makris acknowledged that emissions from the oil fires might cause both acute and chronic health effects and advised that individuals with asthma or chronic obstructive lung diseases be given warnings and precautions. He tempered his comments, however, by saying it was difficult to ascertain long-term health effects "due to insufficient data on the populations exposed, the composition of the smoke plume, the impact of oil pools, and long-term meteorological patterns."

In subsequent hearings before the Senate task force, representatives of other government agencies, including the Public Health Service, consistently cited the failure to resolve the difficulties noted by Makris on the question of public-health risk. In contrast, representatives of private organizations, including Friends of the Earth and the Harvard School of Public Health, urged the environmental and public-health agencies of Kuwait, the United Nations, and the United States to adopt a more aggressive and directed research program. Brent Blackwelder of Friends of the Earth, for instance, cited the failure to perform autopsies on dead livestock and birds.

Meanwhile, environmental researchers from other countries, including Norway and the United Kingdom, arrived in Kuwait, monitored air quality, and released their findings. From this piecemeal effort, observers gradually fit together a coherent, if gap-ridden, picture of Kuwait's environment during the months when the fires yielded to the fire-fighting campaign. Levels of dangerous chemicals in the air were often within the range considered safe by most air-quality standards, but here and there the level of sulfur dioxide exceeded a given standard for a few hours or so. Likewise, the researchers found high levels of polycyclic aromatic hydrocarbons in certain locations for periods of time that varied from minutes to hours. Safety standards for exposure to this class of chemicals are, however, typically defined in terms of yearly averages. The U.S. National Institute of Standards and Technology, the Boston-based National Toxics Campaign, and the U.S. Army Environmental Hygiene Agency all analyzed air samples for chemicals that might lead to cancer. For any given chemical in this class, the researchers only occasionally found levels that might be considered dangerous when sampling at very close range to the fires. But what was the threat posed by inhaling many cancer-causing chemicals at once, even if each particular chemical was present only in minute amounts? Only the toxicological studies came close to addressing this question, and they assessed only the acute threat, not that of cancer.

All groups found high values for particulate matter. The values were especially high if the sampling site stood in the path of a smoke plume, a dust storm, or a combination of the two. Thus, EPA researcher Stevens could say, on the basis of the aggressive sampling program he carried out in July and August, that "Kuwait City looked like a normal city on many days; it was not anything at all like the fires." On other days, however, "the levels of exposure [to pollution from the fires] in Kuwait [were] many times those commonly encountered in urban areas of the United States," according to Evans's testimony to the Senate committee. Drawing attention away from concern about the fires, Stevens reported that Kuwait City "was highly contaminated with sand and auto-exhaust emissions; it had very high concentrations of lead, lead concentrations as high as we used to see in Los Angeles in the early 1970s."

Again and again, at congressional hearings and technical conferences on the fires' environmental consequences, the discussion came back to the basic questions: What contribution did oil-fire smoke make to Kuwait's particle-laden air? How seriously exposed was the population? What fraction of the population was especially sensitive to irritants in the smoke? Without a methodical approach to those topics, no one could assess the risk faced by the population; particulates are a grab bag of whatever is in the air at the time, and if you don't know how much pollution the population is breathing, any effect suffered as a result of pollution will elude you. So with the exception of the handful of scientists who carried out isolated toxicological studies, most of the researchers active in Kuwait during 1991 failed to attack the issue of the particulates' hazy threat in a head-on manner. The limited toxicological studies were the only attempts to make some kind of assessment.

Hanging over the discussions was the specter of the deadly fogs that descended on London during early December in 1952 and 1962. The 1952 fog, which lasted for four days, coincided with about four thousand so-called excess deaths in the London population. Measurements of smoke in

London's air reached values above 4,500 micrograms per cubic meter. Evans included in his Senate testimony a discussion of the London fogs and other incidents in which heavily polluted air coincided with high death rates; he noted that, although the individual risk to the roughly eight million people living in London during the 1952 tragedy was only about one in two thousand, the concern from a public-health perspective was severe.

Comparing the London fogs to the situation in Kuwait is, however, highly dubious. First of all, the air in London was heavily polluted with sulfur dioxide as well as smoke. In 1952, the sulfur dioxide values soared to nearly 4,000 micrograms per cubic meter. According to the British Meteorological Office, the high humidity in London may also have played a role in the fog's deadly power. In addition, more than 80 percent of the Londoners numbered among the excess deaths were over sixty-five; Kuwait's population may have been at less risk, because it is more youthful than London's was in 1952. Nevertheless, in the absence of research that better defined the threat facing Kuwait, George Thurston of New York University and Halûk Özkaynak of Harvard's School of Public Health used data from the London tragedies, as well as from similar incidents in New York and Los Angeles, to estimate the excess mortality Kuwait may have endured. Thurston's calculations showed that about 1,000 excess deaths may have occurred in Kuwait, and Özkaynak estimated that the daily death rate might increase by as much as 10 percent.

The EPA and the U.S. Centers for Disease Control examined the records of emergency rooms at two of Kuwait's six hospitals for the several weeks just before and just after the Iraqis set fire to the oil wells. By determining which complaints were prevalent during the two periods, the government investigators hoped to learn whether the air pollution coincided with more upper respiratory, bronchial, or gastrointestinal complaints and whether they were of an acute or chronic nature. Unfortunately, the population of Kuwait was fluctuating so wildly during the time covered by the survey that the results gave no reasonable indication how an "average" Kuwaiti might have been affected.

Kuwait's Ministry of Public Health has continued to track the correlation between hospital admissions and diagnosis. By late 1991, the ministry announced that more people came to the hospitals and clinics with respiratory complaints, including asthma, during the time when the most wells were burning. But the ministry noted that people were streaming into the hospitals with every sort of complaint then, so that the percentage of complaints that related to breathing difficulty remained about the same throughout 1991. This is difficult information to evaluate. According to Sami al-Yakoob of KEAT, some of the physicians who supplied the records to the ministry are third-country nationals who may have feared for their positions in Kuwait, especially since, at the time, the government was deporting Palestinians and others suspected of disloyalty. If physicians felt intimidated, they might have supplied the ministry with information skewed to support an outcome they believed the ministry was hoping for. For his part, biochemist and KEAT leader Jassem al-Hassan is convinced that asthma and chronic bronchitis sufferers were adversely affected by the smoke from the fires and that government officials did them a poor turn by claiming that the pollutants from the fires were not harmful.

The EPA and Centers for Disease Control have cooperated with Kuwait's Environmental Protection Council in setting up four ongoing studies, which will provide additional information on how Kuwaitis will fare in the aftermath of the crisis and how well the insides of Kuwaiti homes are insulated from the dust that blows about outside. One of these studies will follow the medical histories of 2,000 families, 1,600 of which have members who are asthmatic;

asthma is an important concern in Kuwait; between 8 and 11 percent of the population suffer symptoms of the disease. Another study, which began three weeks before the killing of the last wild well, is comparing the levels of particulates within a sample of Kuwaiti residences to the simultaneous levels of particulates in the air outdoors. By this method, researchers hope to determine how much smoke from the oil fires got into the places where most Kuwaitis were spending most of their time during the crisis.

Even after the last well fires had been put out, it was impossible to give a confident assessment of their effect on human health. The factors that tended to decrease the risk included the relatively small amount of sulfur dioxide and other acidic gases in the plume—in historic incidents of dramatic air pollution, levels of sulfur dioxide and particulates paralleled each other—and the age structure of Kuwait's population. (According to a 1985 census, only 3.5 percent of the population was older than sixty-five, and more than half was under twenty.) It is possible, however, that the low level of sulfur dioxide may not have reduced the health risk, since the toxicological study conducted by the Harvard School of Public Health showed that Kuwaiti air pollution was about as irritating as that in U.S. cities. The contribution of oil-fire smoke to the total amount of solid material floating in the air is equally difficult to assess. Without knowing the times and places where the greatest number of people breathed air laden with smoke particles, researchers will be able neither to demonstrate nor to rule out the suggestion that the fires significantly disturbed the health of people who lived in Kuwait during the crisis.

In a politically charged atmosphere, where the most powerful voices presented a more sanguine view of the public-health threat than the voices of environmentalists and academic researchers did, John Evans outlined work that still could be done to settle the issue. Researchers would have to correlate reliable records of illness and death with day-to-day records of the fire-generated air pollution that the population actually experienced. In his testimony before the Senate task force, Evans noted that the records of pollution exposure would be especially important in such a study, since random errors in measuring the amount of pollution have the effect of hiding cause-and-effect relationships. The Environmental Protection Council's correlation of indoor and outdoor air pollution might prove useful in such a study.

Nevertheless, Evans believes that the best opportunities for epidemiologists to learn from this tragedy were lost because no one collected critical data on the pollution that people experienced at the time when they were coughing and their eyes were running. Neither did anyone identify and study at the time of severest pollution the people in Kuwait who were most likely to suffer because of that pollution.

CHAPTER TEN

The Health Crisis in Iraq

$\Bbb{I}$N STRIKING CON-
trast to the lack of hard information on the suffering and
death the war's environmental effects brought to Kuwait's
population, relief workers and journalists in Iraq docu-
mented a public-health disaster whose human toll was still
rising a year after the cease-fire. Although the well-fire plume
of black smoke often shrouded the entire southern part of
the country, air pollution played only a small role in Iraq's
tragedy. The postwar health crisis centered on water contaminated by infectious diseases, a problem superimposed on
chronic food shortages—which worsened in the year after
the war—in regions of the country where opposition to
Saddam Hussein's regime was strongest.

Kuwait was hardly without its own water and sewage
problem. It was unable to produce any drinking water for
months after the end of the war, and some tank trucks car-

rying water for human consumption contaminated the water with their previous cargo, diesel fuel. Even when the supplies were pure, Kuwaitis complained that their water tasted of smoke, which penetrated into their rooftop water-storage tanks. The war also largely destroyed Kuwait's modern sewage-treatment plant, which had previously purified water to the point where it could safely be used to irrigate food crops. Until the plant's electrical-generating capacity was restored, however, wastewater could not be moved at all, because there was no power for the pumps that sent the sewage to the treatment plant. Once the power came back, people could flush their wastewater away to the plant, but in its ruined state, it could only send the stream on, untreated, into the Gulf. This situation persisted for more than a year after the war. Fortunately, because natural processes similar to composting purify sewage fairly quickly, it did not cause rampant contamination of the Gulf. Treatment plants, however, isolate the process from the wider environment and shrink the amount of space, time, and water necessary for the purification.

In Iraq, the systematic destruction of electric power plants by Coalition air strikes knocked out water-purification and sewage-treatment plants. In the wake of the war, irrigation systems could not function, hospitals were without power, and perishable medicines spoiled. At this point, the two largest segments of Iraqi society—the Kurds of the north and the Shiites of the south, both of which had long resented Saddam Hussein and his Sunni supporters—took encouragement from statements by President Bush and rebelled. They were put down by Hussein within weeks, but more life-sustaining infrastructure was destroyed in the process, along with some of the most sacred shrines of Shiite Muslims. The viciousness with which the Iraqi army turned on citizens of its own country and the horrific images of thousands of Kurdish women and children fleeing from the army over

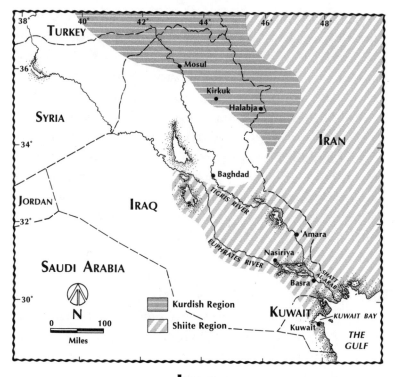

IRAQ

frozen mountain passes forced the United States and the United Kingdom belatedly to send troops to defend the civilian population. Thus, with a second debacle immediately following the Gulf War, Iraq was relegated, in the words of an often-quoted report from the United Nations, "to a pre-industrial age, but with all the disabilities of post-industrial dependency on an intensive use of energy and technology."

At the onset of the war, Iraq was already on the verge of that condition. Beginning almost simultaneously with the Iran-Iraq War in 1980, Saddam Hussein's government implemented a far-ranging plan to transplant large sections of the population from traditional villages that supported themselves largely by agriculture to newly built villages and city neighborhoods, where they would be more dependent on government services and therefore more directly under the control of Saddam Hussein. Some of the new villages housed agricultural workers who continued to work the same land as before. But they had little choice about moving into the concrete apartment houses planted like children's blocks in a sandbox, for Hussein's wrecking crews rapidly demolished the houses in which they and their forebears had lived. The Iraqi government also began an intensive program to industrialize the country with oil-fired electric-power plants. The aim of this policy was not "modernization" or "development" in any benevolent sense of the words but rather the transformation of the country into the military superpower of the Middle East. Chemical plants to manufacture poison gas, biological laboratories to produce agents of germ warfare, and industrial facilities for the construction of nuclear weapons all require prodigious amounts of energy to build and operate. As the new villages and neighborhoods sprang up simultaneously with Hussein's war machine, electrical engineers linked them in the same electric-power grids, effectively making the civilian population hostages to anyone contemplating a wholesale interruption of the weapons-production activities.

By the time the United States and its allies accomplished that interruption in the late February 1991, they had laid waste to Iraq's roads, bridges, communications networks, and electric power. All facilities dependent on electricity, such as hospitals and irrigation pumps, were shut down. In Baghdad and the important industrial city of Basra in the south, citizens had been without regular supplies of fresh water and electricity since the beginning of the air war, in January. The Tigris River, upstream of Baghdad, became an open sewer, as did the Euphrates, and Basra had the unfortunate fate of being located on the combined outfall of the two rivers. The only other ingredient necessary for catastrophe was the displacement of large numbers of people from their homes, and

the Kurdish and Shiite uprisings accomplished that. Greenpeace estimated that, by the time the Iraqi Army had quelled
the uprisings in late April, "well over two million" Iraqis,
or more than 10 percent of the population of nearly nineteen
million, had fled their homes in fear of the army.

In May, a group of legal and public-health specialists
known as the Harvard Study Group issued a report that
seemed to confirm everyone's worst fears. This team spent
nine days in Iraq during late April and early May, visiting
cities in all regions of the country. In its report, it estimated
that, by May 1992, 170,000 Iraqi children under the age of
five would die from "delayed effects of the Gulf Crisis."
Most of the children would die, they predicted, from waterborne infectious diseases like cholera, and many of the projected dead would be weakened by severe malnutrition.

The report had, however, serious flaws. U.N. observers
who visited some of the same sites a few weeks later, for
example, found intact and in limited operation a hospital the
report identified as destroyed. The team used interviews with
physicians at several hospitals to estimate the number of
children who would die from delayed effects of the Gulf
War, and, according to its report, the physicians noted a
doubling to tripling of the death rate in 1991 for young children admitted to those hospitals during the first four months
of the year. Although the team claimed that their projection
of deaths in the year to come was "conservative," many experts considered the report's language inflammatory and overly
pessimistic. The Greenpeace report, which came out at about
the same time, acknowledged that the apocalyptic visions of
postwar Iraq were not supported by most assessments.
Nevertheless, the Harvard Study Group report, however
overstated or imprecise, drew international public attention
to the interconnected problems of electrical-power loss, waterborne infectious diseases, and shortages of food and medicines in the aftermath of a war.

About a month after the Harvard Study Group report, Unicef sent two specialists from the Tufts University School of Nutrition, John O. Field and Robert M. Russell, to Iraq to make a rapid assessment of the nutritional status of children under six years old. They traveled to Baghdad from Amman, Jordan, via the only surface transportation connecting Iraq to the rest of the world—a five-hundred-mile, day-and-a-half taxi ride. In the 110-degree heat of western Iraq, they saw all the wreckage one would expect along one of the air war's more popular strafing runs, but, more significantly from their point of view, they saw a line of trucks bringing food from Amman to Baghdad. "I assume that the food was paid for by some kind of barter system involving the steady stream of trucks carrying oil in the opposite direction," said Russell. Like other visitors who arrived in Baghdad at about that time, Field and Russell were surprised at the seemingly small amount of damage done to residential areas by Coalition air raids. Russell later said that "on the basis of earlier U.N. and Harvard Study Group reports, we could have expected the situation to be much worse."

Field and Russell stayed in Iraq eleven days, eight of them in the southern part of the country, where Unicef believed they would find the Iraqi children most at risk of malnutrition. Allied troops were supervising the distribution of food and medicine among the Kurds in the north, and the food markets of Baghdad had plenty to sell—although at prices three to twenty times higher than before the invasion of Kuwait. But southern Iraq, Basra in particular, was not only the final destination of the untreated sewage flowing down the Tigris and Euphrates rivers but also the far end of the food-distribution line that originated in Baghdad. In every neighborhood of Basra and other cities of the Shiite south, pools of raw sewage cooked in temperatures above 120 degrees Fahrenheit. Typhoid fever, cholera, and other infectious diseases raged through the poor population of the entire region. Mothers traumatized by the events of the previous months were unable to breastfeed their babies. Many of the young children began wasting away, an indication of severe malnutrition known as marasmus. Packs of feral dogs, which had grown used to the taste of human flesh during the Iran-Iraq War, roamed the towns and countryside.

Iraqi doctors don't admit patients to hospitals for malnutrition, and Field and Russell found it remarkable that nutritional deficiencies rarely even figured in diagnoses. Amera Ali, a physician at Ibn Baladi Hospital in Baghdad, was quoted in *The New York Times* as saying: "If we admitted all the marasmus cases, the hospitals would be full in one day." At a hospital in 'Amara, a city on the Tigris, about a hundred miles upstream from Basra, Field and Russell saw a two-year-old girl whose bloated belly showed that she was starving to death; she was in the hospital because she was in the final stages of dehydration brought on by a choleralike disease. Her limbs were wasted by starvation, and her veins too shrunken to admit a needle that might have brought lifesaving liquid. The Iraqi doctors had already done everything possible for her. "Tell the mother that she should stay here with her daughter tonight," Russell said to his interpreter as he moved on. The little girl's knees and elbows were bent at ninety-degree angles, her fists were clenched, her gaping eyes and mouth had frozen into a visage of terror, but no scream, not even a whisper, crossed her thin lips. Earlier, she had screamed—for food, for relief from her cramping insides, for water. Now her throat was too dry to allow the tiniest voice. She lived for another three hours.

She was typical—because the only time in her short life she ever had a chance to fill her tiny stomach was when she was at her mother's breast; because her mother, despite her own hunger, had always managed to get her enough food to keep her alive; because, once the germs that thrive in open cesspools overtook her, there was no clean water to restore her even if the doctors had had the antibiotics to cure her.

Field and Russell visited fourteen communities in the regions around Basra and 'Amara. On their arrival in a community, the local leader, or *mukhtar*, would encourage parents to bring their children to his house, where the Americans weighed and measured them. They examined 680 children in all and found that nearly 40 percent were suffering from chronic malnutrition, as indicated by moderate to severe stunting; 26 percent were moderately to severely underweight; and 26 percent were moderately to severely malnourished, as indicated by the wasting of flesh on the upper arm. They realized they might be witnessing a population on the verge of famine, because, only a month before, a mission from Helen Keller International, an organization dedicated to preventing childhood blindness, had been in the same area and reported figures for upper-arm circumference markedly better than what they found. They concluded, however, that waterborne infectious diseases, spawned in the untreated sewage flowing throughout southern Iraq, posed the greatest threat to public health. Both men emphasized the need for clean water.

Although 78 percent of the mothers who brought infants to them acknowledged that the infants had been sick, usually with diarrhea, in the aftermath of the war and the civil uprising, an even higher proportion—87 percent—of the mothers who came to them with older children told the same story. Field and Russell observed that the slightly better status of the infants could be linked to the breastfeeding habits of Iraqi women, who usually nurse their children for about a year. Although 40 percent of the mothers told the Americans that they could not find essential foods in the markets of southern Iraq, and almost 70 percent said that they could not afford the essential foods that were available, half of the mothers with infants were able to feed the children adequately. The percentage for mothers of older children dropped to less than 40 percent.

Traveling between the region around Baghdad, where the population remained relatively loyal to Saddam Hussein, and the south, which rebelled against him, Field and Russell noted great differences in the degrees to which Iraq had restored not only health services but transportation and electrical power. They credited the Iraqi government with "truly impressive progress in restoring" facilities and services in the region around Baghdad and noted that Iraq was in a better position to "respond effectively to meet nutritional needs and prevent famine" than were other countries with large numbers of hungry people. But they also reported that in the other part of the country, most of the health centers were closed or functioning at a much-reduced scale, despite epidemic levels of infectious disease. At about the same time, news reports out of Iraq claimed that some 80 percent of the country's power grid was still out of service.

In recommending actions for Unicef to take, Field and Russell suggested that the United Nations "will have to decide how to balance the political objectives of the embargo and sanctions against the humanitarian fallout that is virtually certain to follow in their wake." According to Field and Russell, in June 1991 Iraq was not suffering from famine like that in the Horn of Africa, but they cautioned that southern Iraq could face it at any time and urged U.N. officials to take that possibility seriously as long as the Iraqi economy was held back from recovery by United Nations actions. "The situation in southern Iraq could become very much worse very quickly" if Iraq failed to repair more of its sanitary infrastructure as rapidly as possible, they wrote. Such a failure, in combination with continued shortages of food, would put "the health and very lives of hundreds of thousands of children at enormous risk." Given the uncertainty about the U.N. Security Council's resolve to keep in place a trade embargo and sanctions that would constrain economic recovery in Iraq and about the lengths to which Saddam Hussein might take his revenge against disloyal segments of the population, Field and Russell concluded that "anything is possible, including outright disaster, remarkable recovery, even the two together."

As the summer of 1991 wore on, Shiite communities in southern Iraq gradually restored their equipment for disposing of sewage and providing clean water, thus averting the apocalyptic prospect of rampant death from disease and hunger. Restoration came in fits and starts; even after communities had had their electrical power restored, it was not uncommon for it to be down for twelve hours a day or more. As a result, smaller but no less pathetic tragedies abounded. In late June, Patrick Tyler, writing in *The New York Times*, described young children in the advanced stages of marasmus dying by the hundreds in Iraq's hospitals. One humanitarian mission to Iraq after another returned to the West with reports of the suffering endured by Iraq's Kurdish and Shiite populations. "You don't need statistics," Michael Viola of Medicine for Peace was quoted as saying; he saw epidemics of typhoid fever and meningitis "everywhere." The increases in food prices that Field and Russell had noted were conservative compared to figures that surfaced in later accounts. Tyler wrote that, although an infant needed ten cans of powdered formula a month, poor families were allowed to buy only three cans a month at a low, state-subsidized price. For the rest, they had to pay about fifty dollars for a can that had previously gone for just one dollar.

While the United States accused Iraq of blocking the distribution of food and medicine donated by international relief agencies, and Iraqi officials insisted that postwar sanctions were to blame for the travail across the land, a scene that might have been taken for a refugee camp in Kurdistan or the Shiite area of southern Iraq was being played out just south of Iraq's border with Kuwait, in a town named Abdali. There Kuwaiti authorities had interned nearly two thousand Bedu from about three hundred of the stateless Arab families, who commonly work at menial jobs in Kuwait. According to a report by John Cushman in *The New York Times,* the Kuwaitis rounded up the Bedu in northern Kuwait shortly after the end of the Gulf War, claimed that the captured families were of Iraqi descent, and told them to return to Iraq. By July, the camp had already become the birthplace of dozens of Bedu babies. A dishearteningly high proportion of these babies and of others under the age of two were, according to Cushman, suffering from what pediatricians describe as "failure to thrive" syndrome. The disorder often overtakes babies raised under conditions of extreme stress, which apparently characterized the camp at Abdali. In the 120-degree heat of the camp, often under the pall of smoke from the nearby Sabriya and Raudatain fields—the scenes of some of the final and most ghastly fires in Kuwait—and with their parents uncertain about the future, infants stared blankly for hours on end and gave no expression at all of the joy or curiosity that is normal for children their age. The infants' mothers were getting enough to eat and could have been producing enough milk if they themselves had not been so traumatized, but many of the babies had no urge to suckle and wasted away, with diarrhea hastening the process. "Some of the babies seem as malnourished as if they lived in a famine zone," Cushman wrote.

Most of the interned Bedu had fled from Kuwait to Iraq after the invasion by Saddam Hussein's forces, according to Cushman. Others been taken prisoner by the Iraqis, and their families had left Kuwait to find them. After the war, they all hoped to return to their former lives in Kuwait. Months later, Kuwaiti officials finally gave indications that the men who had been taken prisoner originally by the Iraqis would be allowed to return to Kuwait.

Meanwhile, Unicef implemented some of Field and Russell's recommendations, developing a network of nutrition centers across the country to help about 340,000 Iraqi children under the age of six who were believed to be most threatened by severe malnutrition and fatal disease. But the intended beneficiaries, like those who would have been helped by an array of relief institutions, including Catholic Relief Services, have become pawns in the stare-down between Baghdad and Washington over the issue of postwar sanctions. Officials of Saddam Hussein's government, who claim that aid-distribution systems that would operate outside of government control are an infringement of Iraq's sovereignty, disallow such programs in all areas of the country except the Kurd-controlled security zone in the north. Iraq says it is not to blame for the continuing problems of hunger and disease. If the U.N. Security Council would relent on the rules it set down for sales of Iraqi oil—under which profits would be strictly controlled, directed toward humanitarian efforts in Iraq and reparations to Kuwait—Iraq would have sufficient funds to take care of its hunger and medical concerns and would do so, Iraqi officials argue. According to representatives of Unicef and Catholic Relief Services, Iraqi authorities could immediately ease hunger among the country's youngest citizens by releasing millions of dollars' worth of infant formula and high-protein, high-calorie food supplements that are sitting under lock and key in Baghdad and Jordanian warehouses. The U.S. State Department supports this assertion, saying that "it is not the international sanctions which starve Iraq's innocent civilian population, it is the policies of Saddam Hussein's regime." The relief agencies take the position that equitable distribution of already donated food and medical aid would be impossible if carried out under the control of the Baghdad government, a position supported by Jean Mayer, president of Tufts University. Writing on the Op-Ed page of *The New York Times,* he said that "to allow Saddam Hussein and the Government to control relief efforts would consign Iraq's children to a lifetime of malnutrition and neglect."

The number of civilians who have died of hunger and disease has become as politicized an issue as the question of whether the U.N. Security Council should relax sanctions against Iraq in order to speed the delivery of assistance to the afflicted population. In the Security Council, some members have reportedly urged that an embargo against independent deliveries of humanitarian aid, including clothing, school supplies, blankets, and spare parts for electrical water and sewage pumps, be modified to allow such deliveries to Iraq if the donor gives prior notice to the U.N. sanctions committee. The United States is said to have consistently blocked the efforts. Acting as if the death toll of Iraqi civilians from hunger and disease were an embarrassment to the U.S. government, the U.S. Census Bureau tried to fire an analyst who publicized the fact that the bureau had, in its official calculation, clipped her estimate of postwar deaths in Iraq by 10 percent despite at least three prior approvals of the estimate. The analyst, Beth O. Daponte, had originally estimated that Iraqi deaths in the war and its aftermath totaled 86,194 men, 39,612 women, and 32,195 children. A total of 70,000, she believes, died during the twelve months following the war as a result of hunger and disease. In reducing the estimates calculated by Daponte, the bureau concentrated on civilian casualties, dropping her number of 13,000 civilians killed by Coalition action during the war to 5,000. She successfully won an appeal of her dismissal and returned to her position at the bureau.

Epilogue:
The Lessons

⋘ ⋘ ⋘ ⋘ ⋘ ⋘ ⋘ O N T H E F I R S T A N-niversary of Kuwait's liberation, the outline of the Gulf War's environmental disaster was clear. Fueled by a gushing torrent of oil equivalent to about 15 percent of the world's consumption, the fires obscured the sun with a cloud of smoke and petroleum fog that covered a total of some 1.3 million square miles, an area more than twice the size of Alaska. Traces of them dispersed throughout the Northern Hemisphere, possibly contributing to the deaths of hundreds of thousands of people in cyclones and floods. In Kuwait's oil fields, hundreds of millions of barrels of crude oil are likely to remain forever embedded in the desert. The list of catastrophes would have surpassed one's imagination prior to Operation Desert Storm.

There was also the largest—possibly by a factor of more than three—waterborne oil spill in history, leaving hundreds of miles of shoreline slathered in oil more than a foot deep. Despite the greatest cleanup effort in history, in terms of both the actual quantity and the percentage of the spill recovered, the equivalent of four to eight *Exxon Valdez*–sized spills remains trapped and congealing along the shores and bottom of the Gulf. Lighter components of the oil continued to leach back into the Gulf for more than a year after the last of the floating oil was recovered, threatening desalination plants and causing an ongoing irritation to Gulf ecosystems.

The only possible comparisons to this environmental disaster are the catastrophes that struck Bhopal, India, in 1984, killing 1,500 people within three days, and the explosion at Chernobyl, in 1986, which spread deadly cesium 137 throughout the Northern Hemisphere, and which is projected to cause about four hundred deaths by cancer among the 100,000 people who were living within eighteen miles of the nuclear plant. The critical difference is that Bhopal and Chernobyl were accidents—caused perhaps by negligence or wanton disregard for public safety, but accidents nonetheless. In the Gulf War, the stage was deliberately set beforehand, and the instigators had ample information about what would happen. The stage was one of the largest oil reservoirs on earth, set with not only explosive charges on more than seven hundred wells but also the most sophisticated weaponry ever devised. The actors were the forces of Saddam Hussein's Iraq and the United States and its Coalition allies. Millions of civilians in Kuwait, Iraq, Iran, Qatar, Bahrain, and Saudi Arabia were either standing in the wings or had front-row seats. Southern Asians filled most of the rest in the theater. By and large the governments involved preferred to view the spectacle from a respectfully safe distance, by means of remote-sensing technology whenever possible.

Practically everyone I know who was in Kuwait, Iraq, Saudi Arabia, Bahrain, or Qatar between February and August 1991 said that they could never have imagined the all-pervading blackness and stench that filled the sky during those months. Most of them became reflective or sad when I asked them to recall the strongest image from those days. For some, it was the dying moments of a dear friend, a relative, or an anonymous baby. For others, it was the cormorants slowly sinking under a black tide. Some talked about the mythological resonances of the actions of the fire fighters. Everyone agreed that it was a nightmare like nothing the world had ever seen before.

Mohammad Bakr Amin, environmental manager of the Research Institute at Saudi Arabia's King Fahd University of Petroleum and Minerals, told me, "It is a sacred duty for men to prevent pollution and to protect the environment as much as they are able." When he tells his yet-to-be-born grandchildren about the environmental disaster of 1991, he hopes it will make them understand the responsibility that all people bear for taking care of the world they live in. Joe Zoghbi, project manager for Bechtel in Saudi Arabia, told me he will tell his grandchildren that the disaster was a direct result of war. "I want them to know that they should always try to settle their conflicts peacefully," he said, "because the wreckage and danger left behind by belligerency can impose a heavy cost on people for many years."

Long before the fire fighters killed the last oil-well fire, people were discussing how to prevent anything similar from happening again. A good deal of the talk revolved around whether the wording of the 1907 Hague Convention or the 1949 Fourth Geneva Convention encompassed the war's spillage of oil into the Gulf, ignition of hundreds of oil wells, and destruction of essential facilities. The Hague Convention states that belligerents must limit themselves "to those actions that are proportionate to a proper military end," according to the report of the U.S. Senate task force that investigated the environmental aftermath of the Gulf War. The

Geneva Convention bars an occupying power from destroying property except in cases of military necessity. While legal experts who appeared before the task force agreed that Iraq's deliberate spilling of oil and ignition of wells violated those principles, it would be difficult to argue that Saddam Hussein gave much forethought to the consequences of those actions in the context of the conventions.

There are further international agreements that address ecological damage inflicted during, or as acts of, war. Prominent among them are the Geneva Protocols of 1977, which codify the principle that warring states should protect the natural environment from the conduct and effect of their military actions. According to Article 35 of Protocol I, "It is prohibited to employ methods or means of warfare which are intended, or may be expected, to cause widespread, long-term, and severe damage to the natural environment." Neither Iraq nor the United States, however, is a party to the protocol. There is also a U.N. agreement, the Convention on the Prohibition of Military or Any Other Hostile Use of Environmental Modification Techniques, which seeks to limit the manipulation of natural forces for military ends. But its terms are somewhat vague. It does not, for example, prohibit the use of herbicides and other chemical agents. The Senate task force doubted whether Iraq's actions fell within the agreement's regulations.

Amid the differences of opinion on whether Iraq or the United States violated existing international agreements pertaining to environmental devastation inflicted during wartime, some action has been taken that may lead to the adoption of a treaty specifically prohibiting ecological destruction as a weapon of war. Abdullah Toukan, science adviser to King Hussein of Jordan, made the argument for such a treaty in a article published in the Summer 1991 issue of the *Fletcher Forum*. He noted that modern warfare often seeks to destroy facilities that lie outside of "typical military targets" like command, control, and intelligence centers; military air bases; missile sites; tank deployments; artillery; troops; and naval vessels. It also targets facilities on which both civilian and military sectors depend: "oil refineries, oil wells, oil storage tanks, chemical plants, and electric power facilities." Taking the example of the Gulf War, he suggested that the destruction of such civilian facilities, even for military purposes, constitutes using the environment itself as a weapon, since, spread by natural forces, the fallout from a devastating attack on a chemical or nuclear-power plant might lead to widescale contamination that could long outlast the conflict. Toukan's proposal includes a requirement that the environment, "in its broadest sense, must be considered at the initial stages of political and military crisis management." This, he urged, should be a central feature of any treaty designed to supercede the existing U.N. convention on environmental-modification techniques. "This planet and its inhabitants cannot tolerate the shocking waste of human and natural resources caused by the destructive violence of war," his article concludes. "If a new world order is to arise in the aftermath of the Gulf conflict, the protection of the environment must be a central issue."

Toukan observed that the environmental-modification convention of the United Nations was "proven painfully inadequate during the Gulf War"; skeptics may well ask how his proposed treaty would differ. Tyrants bent on imposing their wills across international borders are seldom cowed by international treaties and conventions. Alfred Rubin, professor of international law at Tufts University, argued in a letter to *The New York Times* that discussions over the adequacy of treaties like the 1977 Geneva Protocols serve only to imply that Saddam Hussein did not violate them. "That is to subvert the best safeguard we have for the environment in wartime," he wrote. Better to join ranks and strictly enforce the treaties we already have, he suggested, than to dilute their strength by preparing redundant rules.

International regulation is not, however, our only recourse to prevent environmental disasters. Forward-looking centers of world power could carry out policies that would not only bring economic and environmental benefit to their constituents but also protect people and ecosystems in distant lands. It might be worthwhile to work harder to prevent situations in which we are left waving a piece of paper under the nose of an aggressive despot.

If the industrialized world had not been so dependent on petroleum, Saddam Hussein would have remained as much of an unknown outside the Middle East as Ali Abdullah Saleh, the president of Yemen. Without his income from oil sales, Saddam Hussein could never have developed the vast arsenal he deployed first against Iran during the 1980s and then against Kuwait, the United States, and the rest of the Coalition. In 1989, Iraq exported an average of 2.41 million barrels of oil a day, about half of what Peter Hobbs of the University of Washington calculated was burning in Kuwait during May and June 1991. The export of oil in 1989 earned Saddam Hussein $14.5 billion, which, like his profits from previous years, he plowed into weapons and the development of an arms industry designed to rival that of the United States. The combination of Iraq's petroleum income, Saddam Hussein's freedom to spend the profits however he wanted, and the eagerness of some 450 Western companies to help build Iraq's military-industrial complex created a monster that had to be chastened, but not destroyed, according to Kenneth R. Timmerman. In his book *The Death Lobby*, he writes that foreign companies sent thousands of executives and technicians to Iraq's weapons plants to fulfill lucrative commercial contracts over the course of fifteen years, during which time the United States, its allies, and the International Atomic Energy Agency failed to "hear the truth about Iraq, let alone take the steps to do something about it."

Timmerman documents a shopping spree Saddam Hussein began in France in 1975 and continued almost without a hitch until a few months before the invasion of Kuwait. Just weeks before the invasion, a New Jersey manufacturer of high-temperature furnaces used in the production of nuclear weapons was about to send a shipment to Iraq when President Bush intervened to stop it. It took Iraq's conquest of Kuwait to alert the West to the danger of this market in nuclear technology, a market composed of "bankers, arms salesmen, technology brokers, and Government officials," Timmerman asserts. And he warns that similar dangers are developing in Iran, Syria, and Algeria "for exactly the same reasons."

During 1992, the depth of the United States's involvement in the arming of Saddam Hussein came to light, thanks in part to the work of Texas congressman Henry Gonzalez and his staff. Gonzalez revealed information supporting the view of a questionably cozy relationship between Saddam Hussein and the Reagan and Bush administrations by reading classified documents into the Congressional Record. Among the pieces of this picture is a partially declassified policy directive signed by President Bush in October 1989 which indicated that the United States should "ply Iraq with aid" in order to increase U.S. influence with the country's leader, according to *The New York Times*. Another report told of a covert operation that sidestepped legal restrictions and allowed Saudi Arabia to transfer weapons made in the United States to Saddam Hussein's Iraq. In addition, investigations into a fraud scheme involving billions of dollars at the Atlanta branch of Italy's Banco Nazionale del Lavoro revealed that in 1989 and 1990 the Bush Administration frustrated Federal Reserve Bank officials who were probing Iraq's role in the scandal.

The countries that are home to companies that profit from the sale of militarily significant nuclear technology can do little about the desires of absolute rulers in other parts of the world. But they can be responsible about keeping nuclear-weapons proliferation to a minimum. They can enforce their own regulations of high-technology exports, and they can direct their economies away from dependence on such exports. And they can decrease the value of Iraq's and other oil-exporting countries' assets by promoting greater energy efficiency and developing alternative sources of energy within their own borders.

According to a recent study published by the Union of Concerned Scientists of Cambridge, Massachusetts, if the United States adopted an energy policy that used market forces to encourage more efficient uses of energy as well as to develop renewable and nonpolluting energy sources, it could enhance public health, the environment, and national security, and save money at the same time. Redirecting America's energy policies away from a heavy dependency on imported oil and toward renewable energy sources, as well as emphasizing energy efficiency more in line with the practices of its chief economic rivals—Japan and Europe—would result in "net economic savings amounting to trillions of dollars" over the next forty years.

The United States' burdensome dependency on imported oil—which had reached 42 percent of total consumption by 1990 and is steadily increasing—has played a key role in three major political and economic upheavals since 1972, according to the Union of Concerned Scientists' report. The United States consumes about 26 percent of the world's supply of oil, so any steps the country takes to reduce its oil dependency significantly would simultaneously reduce the influence of oil-exporting countries in geopolitical affairs. The report also notes that imported oil is the single largest component of the United States trade deficit, which retards economic growth and places a heavy burden on Americans of the future. According to the report, the United States could cut its projected primary energy requirements in half over the next forty years, fulfill more than half of those requirements with renewable sources, and more than double the gross national product at the same time.

The energy policies outlined in the report could lead the way to a world less at risk from the possibility of one man's holding the environment hostage and then laying waste to it. Implementing these policies over the next four decades would reduce sulfur-dioxide emissions to 23 percent of the levels projected by the United States government, and reduce nitrogen-oxide emissions to 25 percent of projected levels. Compared to present policy, which would increase carbon-dioxide emissions to more than 50 percent above current levels, new energy policies could lead to almost 30 percent reductions in emissions of the gas most responsible for the greenhouse effect.

A year after the fire fighters in Kuwait battled one of the worst pollution incidents in history, the nations of the world met in Rio de Janeiro at the so-called Earth Summit to agree on plans that would foster global environmental protection while promoting economic development. Increasing numbers of people believe that those two goals are possible, but given the world as it is in 1992, neither is obtainable without efforts leading to the other. The alarming rate of environmental degradation related to declining standards of living in tropical countries is perhaps the best example of the negative aspects of this linkage, aspects that are unfortunately more publicized than the positive. But studies such as that of the Union of Concerned Scientists, and others, illustrate the positive aspects. Faced with this new thinking, some leaders still cling to the old notion of basic conflict between the environment and economic growth and exploit that notion for political advantage. One need only remember the inferno in Kuwait and the smothering death of seabirds in the Gulf to see where that thinking leads.

Source Notes

PROLOGUE—AGAINST THE FIRES OF HELL

p. 3 "He called an official Kuwaiti timetable . . ." "Adair claims Kuwaiti well control may take up to 5 years," *Oil Pollution Bulletin,* June 21, 1991, p. 1.

1—THE FOG OF PEACE

p. 14 ". . . 7,047,000 barrels a day . . ." Figure for March 1991, according to statistics compiled by the American Petroleum Institute from the U.S. Energy Information Administration.

p. 18 "Firefighters claimed they encountered a stubbornness . . ." Matthew L. Wald, "Kuwait Will Hire More Firefighters," *The New York Times,* April 14, 1991.

p. 20 ". . . the air in Kuwait City was never polluted . . ." Interviews with representatives of the Kuwait Oil Company.

p. 21 "Under orders that appeared to have come down from

President Bush . . ." John Horgan, "Up in Flames," *Scientific American,* May 1991.

2—SUCCESS IN THE OIL FIELDS

p. 27 "What this oil's gonna do . . ." Interview, June 3, 1991.

p. 34 "Miller did not deny . . ." Interview, March 4, 1992.

p. 39 ". . . 6.7 million barrels." Average for January and February 1991, according to statistics compiled by the American Petroleum Institute from the U.S. Energy Information Administration.

3—THE BIGGEST OIL SPILL IN HISTORY

p. 45 "Even then, however, a gag order . . ." John Horgan, "Up in Flames," *Scientific American,* May 1991.

p. 46 ". . . a total of 6 million barrels . . ." MEPA communication.

p. 47 ". . . about 2.5 million barrels . . ." MEPA communication.

p. 48 ". . . more than 900,000 barrels . . ." MEPA communication.

p. 48 "About 700,000 barrels . . ." MEPA communication.

p. 50 "In Saudi Arabia, we were coming into a war zone." Interview, March 19, 1991.

p. 52 "The oil spill was without precedent . . ." Interview, November 22, 1991.

p. 54 "The birds we treated . . ." Interview, May 15, 1991.

4—"WE MUST RECOVER THE OIL"

p. 60 "Although al-Gain announced . . ." MEPA press release, February 7, 1991.

p. 63 ". . . Friends of the Earth, which issued a statement . . ." Press release, June 24, 1991.

p. 65 "Every barrel you help us . . ." Interviews with several meeting attendees.

p. 68 "There was nothing else in my mind . . ." Interview, November 23, 1991.

p. 70 "What a lot of people don't realize . . ." Interview, March 19, 1991.

p. 75 "We must recover all the oil . . ." Interviews with participants.

p. 76 "Jim O'Brien viewed the situation . . ." Interview, May 16, 1991.

p. 77 "If that baby works . . ." Interviews with participants.

p. 77 ". . . according to TCA's Rob Lippens . . ." Interview, December 9, 1991.

p. 79 "In Saudi, we didn't have . . ." Interview, May 16, 1991.

5—DIRE PREDICTIONS, SURPRISING MEASUREMENTS

p. 82 "In the 1880s, nearly two hundred refineries operated . . ." Daniel Yergin, *The Prize* (Simon and Schuster: New York, 1991).

p. 84 ". . . where Kuwait and Saudi Arabia share oil production." Figure published by the Kuwait Petroleum Corporation.

p. 84 ". . . accounted for an average 1.59 million barrels a day in 1989." *Oil and Gas Journal,* vol. 88 (1990), pp. 46–83.

p. 85 ". . . 6 percent of the 363 producing wells in Kuwait were classified . . ." OPEC Annual Statistical Bulletin 1989.

p. 87 ". . . would be converted into soot . . ." Richard P. Turco et al., "Climate and Smoke: An Appraisal of Nuclear Winter," *Science,* vol. 247 (January 12, 1990), pp. 166–76.

p. 88 ". . . paper on nuclear-winter theory . . ." Paul J. Crutzen and J. W. Burks, *Ambio,* vol. 11 (1982), p. 114.

p. 90 "Small assumed that . . ." Richard D. Small, "Environmental Impact of Fires in Kuwait," *Nature,* vol. 350 (March 7, 1991), pp. 11–12.

p. 91 ". . . Greenpeace attacked his work . . ." Andre Caroth-ers, "After Desert Storm: The Deluge," *Greenpeace,* October–December 1991, pp. 14–17.

p. 92 "One of the first scientific assessments . . ." S. Bakan et al., "Climate response to smoke from the burning oil wells in Kuwait," *Nature,* vol. 351 (May 30, 1991), pp. 367–71.

p. 93 "A group from the British Meteorological Office . . ." K. A. Browning et al., "Environmental effects from burning oil wells in Kuwait," *Nature,* vol. 351 (May 30, 1991), pp. 363–67.

p. 94 "Friends of the Earth proposed . . ." Press release, June 24, 1991.

p. 95 "On the basis of their sampling . . ." Peter V. Hobbs and Lawrence F. Radke, "Airborne Studies of the Smoke from the Kuwait Oil Fires," *Science,* vol. 256 (May 15, 1992), pp. 987–91.

p. 97 ". . . the carbon dioxide emitted . . ." 21,306 million metric tons of carbon dioxide were emitted by the burning of fossil fuels in 1989. *World Resources 1992–93,* p. 350 (New York: Oxford University Press), 1992.

p. 98 "Imagine all the toxic chemicals . . ." Interview, November 19, 1991.

p. 99 ". . . heaviest rainfall ever recorded in Kuwait . . ." Muhammad Said Subbarini, *Our Marine Environment* (Kuwait: Regional Organization for the Protection of the Marine Environment, 1989).

6—THE SMOKE SEEN ROUND THE WORLD

p. 103 "For the first time in history . . ." Ethan Bronner, "Snow muffles Israeli reply to US on loans," *Boston Globe,* February 26, 1992, p. 20.

p. 104 "In November 1970 . . ." *The World Almanac and Book of Facts 1992,* p. 543 (New York: Pharos Books), 1991.

p. 105 "We lost our ability . . ." Interview, March 31, 1992.

p. 106 "Both events were quite intense . . ." Interview, March 23, 1992.

p. 108 ". . . increases in monsoon precipitation over India." *Report of the WMO Meeting of Experts on the Atmospheric Part of the Joint U.N. Response to the Kuwait Oilfield Fires* (Geneva: World Meteorological Organization, 1991), appendix X, p. 19.

p. 108 "The atmosphere is enormously big . . ." Interview, March 31, 1992.

7—THE TOLL ON PLANTS AND WILDLIFE

p. 110 ". . . he was 'amazed' . . ." Interview, November 23, 1991.

p. 112 ". . . protection of industrial facilities." Meteorology and Environmental Protection Administration, Ministry of Defence and Aviation, Kingdom of Saudi Arabia, *1991 Gulf Oil Spill Shoreline Cleanup Plan* (Jiddah, Saudi Arabia: MEPA, 1991), vol. I, p. 5.

p. 112 ". . . massive damage to the coast." *The Environmental Legacy of the Gulf War* (Amsterdam: Greenpeace International, 1992), p. 16.

p. 114 "The north- and east-facing inlets . . ." Greenpeace video press release.

p. 116 ". . . it was common to find sediments . . ." *Environmental Legacy of the Gulf War*, p. 11.

p. 116 ". . . the situation is very sad . . ." Interview, November 23, 1991.

p. 116 ". . . it's not surprising . . ." Interview, November 23, 1991.

p. 116 "MEPA noted that the low levels . . ." *1991 Gulf Oil Spill*, vol. III.

p. 117 ". . . biologically productive ecosystems of the Saudi coast." Philip W. Basson et al., *Biotypes of the Western Arabian Gulf* (Dhahran, Saudi Arabia: Aramco Department of Loss Prevention and Environmental Affairs, 1977), p. 31.

p. 118 "The use of heavy equipment . . ." Abdulaziz H. Abuzinada, "Towards Restoring the Ecological Viability of the Marine Life in the Gulf." Paper presented at the Conference on Current

Environmental Issues in the Gulf, Dubai, United Arab Emirates, October 22–23, 1991.

p. 122 "This shows that they have been able . . ." Interview, November 22, 1991.

p. 123 ". . . neighboring environments as well." Basson et al., p. 67.

p. 124 ". . . the six-week delay . . ." *Environmental Legacy of the Gulf War*, p. 34.

p. 126 ". . . the absurd cost . . ." Randall Davis and Terrie Williams, Exxon Valdez *Otter Rehabilitation Program* (Galveston: International Wildlife Research, 1990).

p. 126 ". . . disturbance of breeding colonies and oil pollution." *Gulf War Environmental Information Service: Impact on the Marine Environment* (Cambridge, England: World Conservation Monitoring Centre, 1991), p. 18.

p. 128 "We had to take those drastic measures . . ." Interview, November 23, 1991.

p. 128 "It doesn't take much digging . . ." Interview, February 15, 1991.

p. 130 "We want to assist . . ." Interview, November 21, 1991.

p. 130 "LeGore explains . . ." Interview, November 22, 1991.

p. 132 ". . . careful consideration is needed . . ." *Environmental Legacy of the Gulf War*, p. 16.

8—LASTING SCARS

p. 136 "It has been a terrific struggle . . ." Interview, December 2, 1991.

p. 138 ". . . the stones and pebbles effectively still the winds . . ." Interview, March 18, 1992.

p. 140 ". . . the howling winds that blow across the desert . . ." William Branigin, "Kuwait's Economic, Environmental Nightmare," *Washington Post*, March 14, 1991, page A1.

p. 140 "By the end of the dust-storm season . . ." Interview, March 18, 1992.

p. 141 "The increased activity of the dust storms . . ." Interview, March 18, 1992.

p. 142 ". . . amount of motor gasoline burned in California during 1989." According to the U.S. Energy Information Administration, 333 million barrels of motor gasoline were consumed in California during 1989.

p. 143 "Speaking at a conference . . ." Adnan Akbar, "Assessment of the Impact of the Crisis on Ground-Water Pollution from Massive Spillage of Oil from Damaged Wells," *Final Report to Sponsors*, Kuwait Oil Fires Conference, American Academy of Arts and Sciences, Cambridge, Massachusetts, 12–14 August 1991.

p. 148 ". . . the air pollution was one of the most amazing things . . ." Interview, November 12, 1991.

9—BREATHING BLACK AIR

p. 151 ". . . practically every element . . ." Interview, April 13, 1992.

p. 154 "We don't know what the pattern of exposure . . ." Interview, November 28, 1991.

p. 155 ". . . KEAT's appeal was too emotional . . ." Interview, April 21, 1992.

p. 156 "On one occasion . . ." Interview with *New York Times* reporter Matthew Wald, March 7, 1992.

p. 157 "The whole environmental situation . . ." Interview, November 30, 1991.

p. 157 "They went from ignorance . . ." Interview, November 14, 1991.

p. 158 ". . . Matthew Wald reported . . ." "Kuwaitis, Having Survived Hussein, Now Find Their Environment Toxic," *New York Times*, April 28, 1991, p. A14.

p. 159 "Emphasizing the department's concern . . ." Interview, December 2, 1991.

p. 159 ". . . Paul King, who organized the effort . . ." Interview, November 18, 1991.

p. 159 ". . . agreed with King's assessment . . ." Interview, November 18, 1991.

p. 159 ". . . did a magnificent job . . ." Interview, April 13, 1992.

p. 160 "They refused to tell the people . . ." Interview, November 28, 1991.

p. 162 ". . . over twelve hours." Philippe Lameloise and G. Thibaut, AIRPARIF, 10 rue Crillon, Paris 75004, France.

p. 162 ". . . 340 micrograms per cubic meter . . ." U.S. General Accounting Office, "Kuwaiti Oil Fires—Chronic Health Risks Unknown but Assessments Are Under Way," Report GAO/RCED-92-80BR (1992), p. 22.

p. 163 ". . . the team found dangerous chemicals . . ." Jim Makris, testimony before the Gulf Pollution Task Force of the Senate Committee on Environment and Public Works, April 11, 1991.

p. 165 ". . . on the basis of aggressive sampling . . ." Interview, April 13, 1991.

p. 165 ". . . found in the air of U.S. cities." John S. Evans, testimony before the Gulf Pollution Task Force of the Senate Committee on Environment and Public Works, October 16, 1991.

p. 166 ". . . in the absence of research that better defined the threat . . ." Halûk Özkaynak, "Mortality Risks from SO_2 and Particulates: Evidence from New York City and Los Angeles," *Final Report to Sponsors*, Kuwait Oil Fires Conference, American Academy of Arts and Sciences, Cambridge, Massachusetts, 12–14 August 1991.

10—THE HEALTH CRISIS IN IRAQ

p. 172 ". . . Iraq was relegated . . ." Letter dated March 20, 1991 from the secretary-general to the president of the Security Council

(S/22366), enclosing report by Under-Secretary-General Martti Ahtisaari (also known as the Ahtisaari Report).

p. 174 ". . . in fear of the army." William M. Arkin, D. Durrant, and M. Cherni, *On Impact: Modern Warfare and the Environment, a Case Study of the Gulf War* (Washington: Greenpeace, 1991), p. 41.

p. 174 ". . . delayed effects of the Gulf Crisis." Harvard Study Group, "Report on Public Health in Iraq after the Gulf War," p. 1.

p. 174 ". . . by most assessments." Arkin et al., *On Impact,* p. 50.

p. 175 ". . . nutritional status of children . . ." John O. Field and Robert M. Russell, *Nutrition Mission to Iraq: Final Report to Unicef by Tufts University* (1991), p. 46.

p. 176 "Amera Ali, a physician at Ibn Baladi Hospital . . ." Patrick E. Tyler, "Disease Spirals in Iraq as Embargo Takes Its Toll," *The New York Times,* June 24, 1991, p. A1.

p. 179 "In late June . . ." Tyler, "Disease Spirals."

p. 180 "According to a report . . ." John Cushman, "The Babies That Won't Blossom," *The New York Times,* July 16, 1991, p. A3.

p. 181 ". . . not the international sanctions . . ." Patrick E. Tyler, "U.S. Accuses Iraq of Blocking Aid," *The New York Times,* January 4, 1992, p. A1.

p. 182 ". . . to allow Saddam Hussein . . ." Jean Mayer, "Iraq's Malnourished Children," *The New York Times,* July 15, 1991.

EPILOGUE—THE LESSONS

p. 184 ". . . within eighteen miles of the nuclear plant." U.S. Department of Energy figure, cited in *World Resources 1988–89* (New York: Basic Books, 1988), p. 46.

p. 185 ". . . environmental aftermath of the Gulf War." United States Senate Committee on Environment and Public Works Gulf

Pollution Task Force, *The Environmental Aftermath of the Gulf War* (Washington: Congressional Research Service, 1992), p. iv.

p. 187 ". . . the environment must be a central issue." Abdullah Toukan, "The Gulf War and the Environment: The Need for a Treaty Prohibiting Ecological Destruction as a Weapon of War," *Fletcher Forum*, Summer 1991, pp. 95–100.

p. 187 "That is to subvert . . ." *The New York Times*, July 1, 1991, p. A12.

p. 188 ". . . chastened, but not destroyed . . ." Kenneth R. Timmerman, *The Death Lobby* (Boston: Houghton Mifflin), 1991.

p. 190 ". . . over the next forty years." *America's Energy Choices: Investing in a Strong Economy and a Clean Environment* (Cambridge, Mass.: Union of Concerned Scientists, 1991).

Index

203